Medical Applications of Fluorescent Excitation Analysis

Editors:

Leon Kaufman
Head, Experimental Nuclear Instrumentation
Radiologic Imaging Laboratory
University of California
San Francisco, California

David C. Price
Head, Section of Nuclear Medicine
Department of Radiology
University of California
San Francisco, California

CRC Press
Taylor & Francis Group
Boca Raton London New York

CRC Press is an imprint of the
Taylor & Francis Group, an **informa** business

First published 1979 by CRC Press
Taylor & Francis Group
6000 Broken Sound Parkway NW, Suite 300
Boca Raton, FL 33487-2742

Reissued 2018 by CRC Press

Library of Congress Cataloging in Publication Data

Main entry under title:

Medical applications of fluorescent excitation analysis.

Bibliography: p.
Includes index.
1. X-ray spectroscopy. 2. X-ray spectroscopy- -
Instruments. 3. Diagnosis, Laboratory. I.
Kaufman, Leon. II. Price, David C.
QP519.9.Z73M42 616.07'57 78-27604
ISBN 0-8493-5507-9

A Library of Congress record exists under LC control number: 78027604

ISBN 13: 978-1-315-89526-0 (hbk)
ISBN 13: 978-1-351-07436-0 (ebk)

Visit the Taylor & Francis Web site at http://www.taylorandfrancis.com and the
CRC Press Web site at http://www.crcpress.com

PREFACE

Fluorescent excitation analysis (FEA) is a technique that has been utilized for some time in physics. An increasing number of biomedical applications for FEA have been reported in recent years; it is becoming the assay method of choice in many areas of research and clinical practice. The excellent developmental work continuing in various medical centers and federal laboratories has also resulted in the availability of high-quality equipment for FEA, including equipment developed specifically with the biomedical market as its primary target.

The purpose of this volume is to acquaint the interested physician or physicist with the basic principles and the instrumentation relevant to FEA, as well as some present and future biomedical applications. Only such a comprehensive understanding of FEA will permit rational selection of equipment according to its performance specifications, and continuing development of new procedures that effectively utilize the advantages provided by FEA.

Leon Kaufman
David C. Price

THE EDITORS

Leon Kaufman, Ph.D., is Director of the UCSF Radiologic Imaging Laboratory and Head of Experimental Nuclear Instrumentation in the Department of Radiology at the University of California, San Francisco, where he holds an appointment as Professor of Physics. Dr. Kaufman received his B.S. degree in Engineering Physics in 1964 and Ph.D. in Physics in 1967, both from the University of California at Berkeley. He joined the Section of Nuclear Medicine, Department of Radiology, in 1970. His research has covered, in addition to fluorescent excitation analysis, the development of new gamma and X-ray imaging techniques for non-invasive diagnosis, including new display systems, the use of multiwire proportional chambers, high-purity germanium cameras, room temperature semiconductor detectors, nuclear magnetic resonance and positron tomography. Dr. Kaufman has served in the IEEE, the American Nuclear Society and is a past president of the Northern California Chapter of the Society of Nuclear Medicine. Dr. Kaufman has published in the field of high energy and nuclear physics, in addition to having contributed many publications in the field of radiology.

David C. Price received his M.D. degree from the University of Toronto, then pursued graduate training in Internal Medicine in Toronto and Vancouver. During this time he spent two years at the Medical Research Center, Brookhaven National Laboratory, gaining clinical and research experience in Hematology and in the beginning field of Nuclear Medicine. After three years of additional training at Donner Laboratory, Lawrence Berkely Laboratory, he joined the faculty of the School of Medicine at the University of California at San Francisco. In 1977 he was appointed Head of the Nuclear Medicine Section, Department of Radiology. In addition to clinical, teaching, and administrative responsibilities, Dr. Price maintains research interests in new instrumentation (such as fluorescent excitation analysis) and in Nuclear Hematology, including body compartment studies, cell survival kinetics and bone marrow scintigraphy.

CONTRIBUTORS

David C. Camp
Senior Scientist
Nuclear Chemistry Division
Lawrence Livermore Laboratory
Livermore, California

Frank Deconinck
Vrije Universiteit Brussel
Dienst Radioisotopen
Brussels, Belgium

Fred S. Goulding
Head, Department of Instrument Techniques
Lawrence Berkeley Laboratory
University of California
Berkeley, California

Joseph M. Jaklevic
Physicist, Group Leader
Department of Instrument Techniques
Lawrence Berkeley Laboratory
University of California
Berkeley, California

Leon Kaufman
Head, Experimental Nuclear Instrumentation
Radiologic Imaging Laboratory
University of California
San Francisco, California

James A. Nelson
Associate Professor
Department of Radiology
University of Utah Medical Center
Salt Lake City, Utah 84112

Michael D. Okerlund
Associate Professor
Nuclear Medicine Section
Department of Radiology
University of California Medical Center
San Francisco, California

P. S. Ong
Associate Professor
University of Houston
Electrical Engineering Department
Houston, Texas 77004

David W. Palmer
Assistant Professor
Milwaukee County Medical Complex
Division of Nuclear Medicine
Milwaukee, Wisconsin

David C. Price
Head, Section of Nuclear Medicine
Department of Radiology
University of California
San Francisco, California

TABLE OF CONTENTS

Section I — Principles

Chapter 1
Physical Principles 3
David C. Camp

Chapter 2
Instrumentation for Energy Dispersive X-Ray Fluorescence 29
Joseph M. Jaklevic and Fred S. Goulding

Chapter 3
Excitation Methods for Energy Dispersive Analysis 49
Joseph M. Jaklevic

Section II — Trace Elements

Chapter 4
Trace Elements in Medicine 71
P. S. Ong

Section III — Stable Tracers

Chapter 5
Techniques for In Vitro Fluorescent Excitation Analysis of Stable Tracers 91
Leon Kaufman

Chapter 6
Fluorescent Excitation Analysis in the Measurement of Body Water Compartments, Cell Survival, and Specific Organ Function 115
David C. Price

Chapter 7
Studies of the Kinetics of X-Ray Contrast Agents Using Fluorescent Excitation Analysis 129
James A. Nelson

Section IV — Thyroid Studies

Chapter 8
Current Developments in Instrumentation for Fluorescent Scanning of the Thyroid 139
David W. Palmer, Leon Kaufman, and Frank Deconick

Chapter 9
The Clinical Utility of Fluorescent Scanning of the Thyroid 149
Micahel D. Okerlund

Index 163

Section I

Principles

Chapter 1

PHYSICAL PRINCIPLES

David C. Camp

TABLE OF CONTENTS

I. Introduction 3
A. Historical Development 3
B. Overview of Fluorescent Excitation Analysis 4
C. Advantages and Applications of FEA 5

II. Atomic Structure 6
A. The Nuclear Atom 6
B. X-Ray Nomenclature and Energies 7
C. X-Ray Intensities and Fluorescence Yields 12

III. Fluorescent Excitation Analysis 14
A. Radiation Absorption and Attenuation 14
B. Photoelectric Absorption 17
C. Coherent Radiation Scattering 19
D. Incoherent Radiation Scattering 19
E. Polarized Radiation 20
F. Summary 22

References 27

I. INTRODUCTION

A. Historical Development

Within 1 year following Roentgen's[1] discovery of X-rays in 1895, they were being applied in medical and industrial radiography and fluoroscopy. In 1913, Moseley[2] demonstrated a well-ordered relationship between X-ray energy and atomic element; and in this same paper demonstrated that X-rays were a useful tool for both qualitative and quantitative spectrochemical analysis. In the same year, Bohr[3] published his interpretation of the atomic spectrum of the hydrogen atom. He introduced the idea that the motion of electrons around the nucleus was restricted or quantized, and that they could exist only in certain orbits or energy levels. Thus, when an atomic transition occurred, an atomic electron changed from one energy level to another, and energy was released only in fixed and predictable amounts. The applicability of Bohr's theory to Moseley's results was recognized at once, and was a great triumph for this new quantum theory. Less than 10 years later, stimulated X-ray emission from minerals was used[4] to determine their elemental composition and abundances.

Much of early X-ray spectrochemical analysis was carried out using photographic emulsions and crystal diffractometers. Commercial secondary emission and electron probe X-ray spectrometers were not developed until the late 1940s. They employed gas proportional counters, and somewhat later, scintillation counters to detect the X-rays.

The modern and widespread use of energy-dispersive X-ray spectrometry, as opposed to wavelength-dispersive spectrometry using diffractometers and goniometers, had to await development of semiconductor radiation detectors, particularly the lithium-drifted silicon detector, and solid-state, pulse-height analyzers. The fundamental principles and important operational characteristics of these two are discussed in more detail in Chapter 2. Modern energy-dispersive X-ray spectrometry is usually aknowledged to have begun in 1966 when Bowman et al.[5] first demonstrated the analytical capabilities of a semiconductor detector. Since that time, many different terms for this kind of spectrochemical analysis have appeared in the literature. These include X-ray emission spectroscopy (XES), X-ray fluorescence analysis (XRFA), energy dispersive X-ray fluorescence analysis (EDXRFA), and fluorescent excitation analysis (FEA). Throughout this work, the latter term will be used.

B. Overview of Fluorescent Excitation Analysis

The analytical technique of fluorescent-excitation analysis, or FEA, depends upon the ability to excite a sample and accurately measure the characteristic X-rays emitted from the excited atoms. There are many different ways to excite atoms within a sample. These include the use of ordinary low- or high-power X-ray tubes, with or without filtered anodes; irradiation by α-, β-, γ-, or X-rays from radioisotopes; bombardment by charged particles or ions from particle accelerators; bombardment by electrons, as in electron microprobes; irradiation by secondary X-rays from a selected target element, or by polarized X-rays such as from synchrotron radiation; self-excitation, if the sample contains some radioactivity; and by observing those X-rays which follow certain nuclear decay modes. The principles, practical sensitivities, and usefulness of some of these methods are discussed in Chapter 3.

In order to understand and interpret spectral distributions and X-ray lines as displayed on modern multichannel analyzers, it is necessary to gain some understanding of atomic structure and of the radiation absorbtion and scattering processes that lead to the observed spectrum. Each of these, in turn, will be discussed in Sections II and III of this Chapter. However, before treating these topics in detail, a brief description of the entire FEA process will be helpful.

Assume that a flux of exciting radiation composed of particles or photons of energy, E_p, is incident on a sample. The fundamental unit of energy is the electron volt; that is, the amount of energy gained by an electron falling through a potential difference of 1 V. Since this is a very small amount of energy and since most X-rays have thousands of electron volts of energy, the kilo-electron volt (keV) unit is more useful. A small part of the incident radiation flux, depending on the sample's thickness and on the energy of the exciting radiation, may not interact with the sample at all. Another part of the incident flux may scatter, either with or without some energy loss; or part of the flux may be completely absorbed by the sample. Although there may be millions of each of these interactions throughout the sample, each interaction must be thought of as taking place with a single atom. A simplified representation of one such interaction is illustrated in Figure 1. If the incident quantum is totally absorbed and if E_p(keV) is greater than the binding energy of some electron in the atom, then one of the atom's shell electrons will be ejected. This creates a vacancy in that atomic shell, which leaves the atom in an excited or unstable condition. The vacancy is filled by an atomic transition, that is, by an electron which falls from one of the less tightly bound atomic shells into the inner shell vacancy. This results in either the emission of one of the atom's characteristic X-rays, or an Auger electron (which rarely escapes the sample). Thus, an X-ray is not always emitted following atomic excitation (quantum absorbtion). Once an X-ray is emitted from the atom and escapes the sample, it is available for spectroscopic analysis. Since all of the characteristic X-rays associated with any

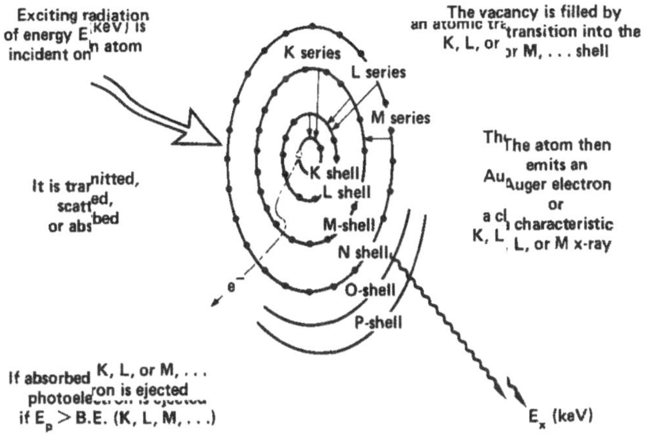

FIGURE 1. A simplified schematic representation of X-ray fluorescent excitation analysis. The number of electrons within a given shell and the number of major shells depend on the particular atom. B.E. stands for binding energy.

element are well known, it is possible to identify, qualitatively, most of the elemental constituents of a sample through fluorescent excitation analysis. Quantitative analysis requires that considerably more attention be paid to several experimental parameters. These include sample preparation, the experimental apparatus and its associated geometries, the preparation of standards for calibration, and any attenuation of X-rays from self-absorption within the sample.

C. Advantages and Applications of FEA

Rarely has any new analytical technique spread so quickly into so many different professional disciplines. The principal reasons for this are found in the technique's advantages. FEA offers a very wide dynamic range in elemental concentrations that may be detected, from constituents in the percent range down to trace quantities in the parts-per-billion range. Quantitatively, the technique is rapid, offering fractional second identification for on-line manufacturing processes, to minutes for routine sample analysis. Electron or photon excitation is nondestructive to a sample, while charged particle beams from accelerators may be destructive. Compared to classical analytical chemistry techniques, which are destructive to a sample and require careful and elaborate sample preparation, FEA requires little to no sample preparation. The sample form may be solid, liquid, or even gaseous. Normally, from one to three analyses will yield all of the elemental constituents above sodium within the sample; thus, the technique is often referred to as multi-element. For elements below sodium, the emitted X-ray energies are very soft (or low in energy) and are usually self-absorbed in the sample unless very special sample preparation and analysis procedures are used. The spectral distribution of X-rays resulting from FEA is almost always unambiguous. The few cases where ambiguity does occur are well known by experienced analysts. In most FEA spectroscopy systems multichannel analyzers are essential. The recorded data are stored in digital form and, therefore, easily routed to and manipulated by microprocessors or minicomputers. Once standard sample geometries or containers are adopted, the analysis of many such samples can be easily automated, thus making the cost per sample inexpensive. Finally, the equipment associated with FEA spectroscopy, even if automated, is simple to understand and operate; thus these tasks can be performed by trained technicians. Clearly, one or several of the above advantages are offered by other analytical techniques, but when all of them are taken together, they demonstrate why FEA has enjoyed such rapid and wide acceptance.

The many advantages offered by FEA has brought about a rather broad range of applications. It has been used in the trace element analysis of aerosol particulate matter,[6] in the characterization of other environmental pollutants found in urban and rural atmospheres,[7] and in the analysis of samples gathered near coal-fired power plants.[8] It is easily adapted to the analysis of commercial products such as cement,[9] or industrial metals for selected impurities;[10] and it is used for the in-stream analysis of ore-grade slurries in mining operations.[11] An analysis and cataloguing of the trace element composition of selected paint colors as a function of artistry has enabled FEA to distinguish art forgeries from originals.[12] Similarly, the technique can be applied to coins, sculptures, ceramics, and other art objects,[13] as well as pottery fragments[14] and similar archeological antiquities.[15] It is also useful in the fields of criminology[16] and geology, including a simple analysis for the major constituents of lunar[17] and planetary surfaces.[18] Finally, it has been applied to the trace-element analysis of many types of medical samples. Some of these will be discussed in Chapter 4. Two other applications of FEA, i.e., the use of stable tracers for disease diagnosis and for thyroid scanning, are discussed in Chapters 5 and 6. Clearly, FEA is a versatile analytical technique.

II. ATOMIC STRUCTURE

A. The Nuclear Atom

The spectroscopic and chemical properties of the elements are almost entirely determined by the atomic number, Z, which gives the integral number of protons within the nucleus. This nuclear charge governs the number of atomic electrons which can surround the nucleus. The arrangement of atoms according to increasing atomic number is useful to spectroscopists, while the chemist classifies them into chemical elements and groups of elements, hence, the Periodic Table of the Elements (see Appendix 1). All elements, stable or unstable, can exist in the form of different isotopes, that is atoms having the same atomic number or charge (Z), but a different number of neutrons, N. The total number of neutrons and protons, or nucleons, within the nucleus is called the mass number, A.

Elements occur in nature for almost all integral nuclear charge up to $Z = 92$, uranium. Exceptions occur for $Z = 43$, technetium, which has no stable isotope; and for the elements $Z = 84$ to 91, which also have no stable isotopes, but can be found in nature because they are radioactive decay products of very long lived uranium and thorium isotopes. Substantial to small amounts of the elements from $Z = 93$, neptunium, to $Z = 100$, fermium, have been made in particle accelerators and nuclear reactors and their chemical properties studied, but beyond, from $Z = 101$ to 106, a few million to only a few hundred atoms, respectively, have been made. No well accepted evidence exists to date (1978) for any element in the so-called superheavy region, $Z =$ 110 to 130. There are a number of radioactive isotopes useful for FEA, which will be introduced in Chapter 3. The most complete reference on stable and unstable isotopes, their properties, and any associated radioactive decay modes is published in the Table of Isotopes.[19]

The nuclear model of the atom proposed by Bohr, and later developed more fully by quantum mechanics, results in a distribution of electrons around the atom being confined to specific quantized levels or shells and even subshells. The negatively charged electrons are bound to the positively charged nucleus containing Z protons by the attractive Coulomb force. Each electron bound within any shell or subshell will have a well-characterized set of four quantum numbers, no two of which can be the same (Pauli exclusion principle). Each of these quantum numbers describes an important property of the atomic electron, and they arise quite naturally from the mathematical theory which describes the atom. The principal quantum number, $n = 1, 2,$

..., defines the major shell. The quantum number, l = 0, 1, 2, ... (n-1), is related to the electron's orbital angular momentum. A quantum number, $j = l \pm \frac{1}{2}$, describes the vector sum of l and s, the intrinsic-spin quantum number, which has values $\pm \frac{1}{2}$ (corresponding to a spin up or clockwise motion, or spin down, a counter-clockwise motion). Since atomic transitions, including X-rays, correspond to electron transfers from one shell to another, some use of these quantum numbers is required in order to understand the X-ray nomenclature that corresponds to the various atomic transitions.

The most tightly bound shell is assigned the principal quantum number, $n = 1$. This shell can have only two electrons in it (spin up and spin down), and in X-ray terminology, is referred to as the K shell. If an electron vacancy is created in this shell, the atom will be in its highest state of single-electron excitation, and an electron which falls from any less tightly bound shell into this vacancy will give rise to X-rays belonging to the K series. Transitions involving no change in orbital angular momentum are forbidden.

The next most tightly bound shell is the $n = 2$, or L shell, which is composed of three subshells: $n = 2, l = 0, j = \frac{1}{2}$, the LI shell; $n = 2, l = 1, j = (l-s) = \frac{1}{2}$, the LII shell, and $n = 2, l = 1, j = (l+s) = 3/2$, the LIII shell. The maximum electron population allowed for any subshell is given by $2j+1$ and for any entire shell by $2n^2$. All electron transitions from any less tightly bound subshell to a vacancy within one of the L subshells are known as L-series X-rays.

Similarly for the n = 3, 4, 5, ... (M, N, O, ...) shells, the various subshell configuration and electron populations can be built up. A few atomic transitions corresponding to K-, L-, and M-series X-rays were illustrated in Figure 1, although no subshell structure was shown. Clearly, the orbital electronic structure of any atom depends on the nuclear charge. The light elements have very little nuclear charge, hence few orbital electrons, thus only K-series X-rays exist. On the other hand, for the heavier elements there are many shells and subshells and with the proper experimental equipment, all of the possible X-ray series can be observed. The number of electrons in each major shell for each element is shown on the Periodic Table of Elements, Appendix 1.

B. X-Ray Nomenclature and Energies

Unfortunately, X-ray nomenclature has evolved historically so that there is no simple relationship between X-ray designation and the corresponding atomic transition. In FEA applications, it is not the atomic origin that is crucial, but rather the X-ray energy that is the most important quantity leading to its identification. As noted in the previous section, all K-series X-rays correspond to the filling of a K-shell vacancy, L-series X-rays to the filling of L subshell vacancies, and so forth. In general, a greek letter attached to an X-ray indicates its relative intensity. That is $K\alpha$ is stronger than $K\beta$; while $L\alpha$ is more intense than $L\beta$. X-ray intensities are discussed in the next section.

Table 1 lists the X-ray nomenclature currently associated with selected atomic transitions. The $K\alpha 1$ and $K\alpha 2$ X-rays correspond to electron transitions from the LIII and LII subshells to the K shell, respectively. The $K\beta$ X-rays with odd order correspond to electron transitions from the M subshells, while those with even order originate from N subshells. The $K\delta$ X-rays are weak and can be observed with energy dispersive systems only for high Z ($Z > 60$).

The L series of X-rays correspond to the filling of vacancies in the LI, LII, and LIII subshells. From the table it can be seen that $L\alpha 1$ and $L\alpha 2$ X-rays correspond to the filling of LIII subshell vacancies by electrons from the MV and MIV subshells, respectively; while $L\beta 1$ and $L\beta 2$ correspond to LII and LIII vacancies being filled by MIV and NV subshell electrons, respectively. Note that there are many more L-series X-rays than K-series X-rays. For the heavy elements, such as lead ($Z = 82$) or uranium ($Z = 92$), many of these L X-rays are energetic enough to be observed easily in FEA.

TABLE 1

X-ray Nomenclature vs. Atomic Transition

Initial level[a]				Final level[a] in the						
nℓ	Shell	j	#e⁻	K-,	L-,		or	M-Shell		
1 s	K	1/2	2	K Series						
2 s	LI	1/2	2			L Series[b]				
2 p	LII	1/2	2	Kα2						
2 p	LIII	3/2	4	Kα1	LI	LII	LIII			
3 s	MI	1/2	2			Lη	Lℓ			
3 p	MII	1/2	2	Kβ3	Lβ4		Lt			
3 p	MIII	3/2	4	Kβ3	Lβ3	Lβ17	Ls		M Series[c]	
3 d	MIV	3/2	4	Kβ5″	Lβ10	Lβ1	Lα2			
3 d	MV	5/2	6	Kβ5′	Lβ9		Lα1	MIII	MIV	MV
4 s	NI	1/2	2			Lγ5	Lβ6			
4 p	NII	1/2	2	Kβ2″	Lγ2					
4 p	NIII	3/2	4	Kβ2′	Lγ3				Mζ2	Mζ1
4 d	NIV	3/2	4			Lγ1	Lβ15	Mγ2		
				Kβ4						
4 d	NV	5/2	6		Lγ11		Lβ2	Mγ1		
4 f	NVI	5/2	6			Lv			Mβ1	Mα2
							Lu			
4 f	NVII	7/2	8							Mα1
5 s	OI[d]	1/2	2			Lγ8	Lβ7			
5 p	OII	1/2	2	Kδ2	Lγ4′					
5 p	OIII	3/2	4	Kδ1	Lγ4					
							Lβ5			
5 d	OIV	3/2	4			Lγ6				
5 d	OV	5/2	6					Mε		
5 f	OVI	5/2	6							
5 f	OVII	7/2	8							
5 g	OVIII	7/2	8							
5 g	OIX	9/2	10							

[a] Initial and Final levels refer to the electron's position.

[b] Most of the assigned L-series X-rays are identified here.

[c] Only a very few M-series X-rays are identified using greek letters.

[d] the O shell begins to fill for Z ⩾ 37, rubidium; the P shell for Z ⩾ 55, cesium. The N series of X-rays are rarely observed and have very low energies (<1 keV), even for very heavy elements.

Only a few of the M-series X-rays are assigned greek letters. For all but the heaviest elements, the M-series X-rays have very low energies and intensities and are only occasionally observed in FEA. For those X-rays of any series K, L, M, N, ..., which do not already have an assigned nomenclature, the initial and final electron locations involved in the atomic transition can be used to identify them, i.e., (MV → LIII) is assigned as Lα1, but the M-series X-ray (0VII → MV) does not have any associated symbolism.

For all atoms, the K-shell electrons are more tightly bound than the L-shell electrons, and those electrons more so than the M-shell electrons. The greatest change in energy that an electron can undergo will occur between the outermost electron shell and the innermost K shell; while the least change will occur for a transition between the two outermost shells. As the atomic number or nuclear charge increases the electrons in all shells are bound more tightly. Thus, the more charge the nucleus has, the greater the energy release for atomic transitions between the same two subshells. Figure 2 shows how the energy of the Kα and Kβ X-rays (and their energy difference) increases with increasing Z. The L X-ray energies also increase with Z, but have only about one-sixth the energy of the K X-rays for a given Z. Although it is not readily apparent on the small scale of Figure 2, the separation between the Kα1 and Kα2 and the Kβ1 and Kβ2 components also increases with Z. Since the energy of any X-ray is the single most important parameter leading to its identification, the energies of the stronger X-rays must be known accurately. Table 2 lists X-ray energies to the nearest electron volt for the elements Z = 1 to 100. Those X-rays listed are Kα1, Kα2, Kβ1, and Kβ2 of the K series; Lα1, Lα2, Lβ1, Lβ2, Lβ3, Lγ1, Lγ3, Lℓ, and Lη of the L series; and Mα1, Mα2, and Mβ1 of the M series.

In most FEA applications, the K X-rays will be the most useful in both qualitative and quantitative analysis. From Table 2, it is seen that the energy separation between the Kα and Kβ X-rays for Z > 18 (argon) becomes great enough so that FEA spectroscopy systems having a resolution of 160 eV FWHM* at 3.0 keV will begin to resolve these two components. The Kβ1 and Kβ2 X-rays begin to be resolved for X > 40 (zirconium) and Kα1 begins to be clearly separated from Kα2 above Z = 60 (neodymium). For the heavier elements (Z = 82 ± 10), the L series of X-rays have energies in the 8 to 20-keV range and can become as useful as the K series for qualitative and quantitative analysis. Only in special applications will M X-rays be encountered. Even for lead (Z = 82), the mean energy of the Mα and Mβ X-rays is only 2.39 keV. One occasion

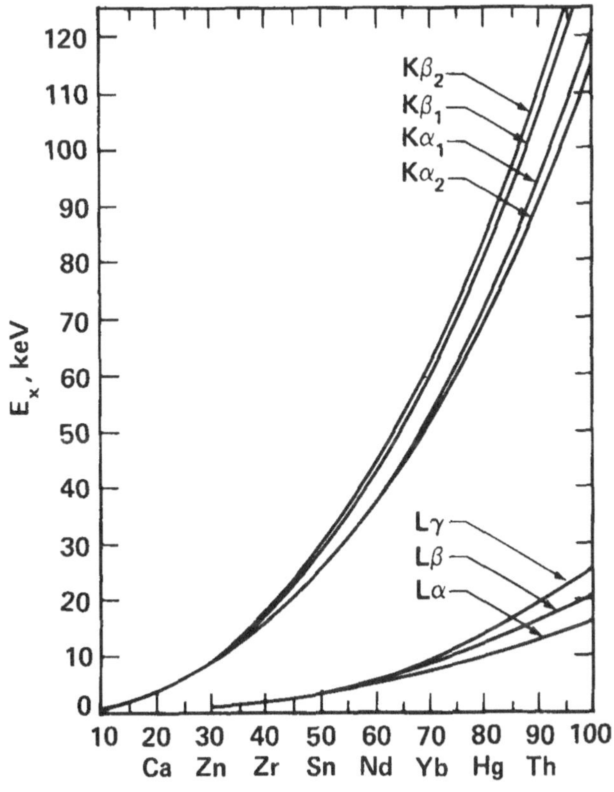

FIGURE 2. Energies of the K and L series of X-rays as a function of atomic number. Note that the energy separation between the Kα and Kβ and between the three L components increases with Z.

* FWHM ≡ The Full Width of the X-ray peak at Half the peak's Maximum height. See Chapter 2.

TABLE 2

Absorption Edge and X-Ray Energies

	K SERIES					L SERIES												
Z Element	Kab	Kα1	Kα2	Kβ1	Kβ2	LIab	LIIab	LIIIab	Lα1	Lα2	Lβ1	Lβ2	Lβ3	Lγ1	Lγ3	Lℓ	Lη	Z Element
1 H	0.0136																	
2 He	0.025																	
3 Li	0.055	0.052																
4 Be	0.112	0.110																
5 B	0.192	0.185																
6 C	0.283	0.277																
7 N	0.399	0.392																
8 O	0.531	0.525																
9 F	0.687	0.677																
10 Ne	0.867	0.848																
11 Na	1.072	1.041		1.067														
12 Mg	1.305	1.253		1.295														
13 Al	1.559	1.486	1.486	1.553														
14 Si	1.838	1.740	1.739	1.829														
15 P	2.142	2.013	2.012	2.136														
16 S	2.472	2.307	2.306	2.464														
17 Cl	2.822	2.622	2.620	2.815														
18 Ar	3.202	2.957	2.955	3.190														
19 K	3.607	3.313	3.310	3.589												0.260	0.262	K 19
20 Ca	4.038	3.691	3.687	4.012		0.400	0.350	0.346	0.341		0.345			0.350		0.303	0.306	Ca 20
21 Sc	4.496	4.090	4.085	4.460		0.463	0.407	0.403	0.395		0.400			0.407		0.348	0.353	Sc 21
22 Ti	4.965	4.510	4.504	4.931		0.530	0.461	0.454	0.452		0.458			0.460		0.395	0.401	Ti 22
23 V	5.465	4.951	4.944	5.426		0.604	0.520	0.513	0.511		0.519		0.585	0.520		0.446	0.453	V 23
24 Cr	5.989	5.414	5.405	5.946		0.682	0.584	0.574	0.573		0.583		0.654	0.583		0.500	0.510	Cr 24
25 Mn	6.540	5.898	5.887	6.489		0.754	0.652	0.641	0.637		0.649		0.721	0.652		0.556	0.567	Mn 25
26 Fe	7.112	6.403	6.390	7.057		0.842	0.723	0.709	0.705		0.718		0.792	0.721		0.615	0.628	Fe 26
27 Co	7.709	6.929	6.914	7.648		0.929	0.794	0.779	0.776		0.791		0.866	0.794		0.678	0.694	Co 27
28 Ni	8.333	7.477	7.460	8.263		1.012	0.871	0.855	0.851		0.869		0.941	0.872		0.743	0.762	Ni 28
29 Cu	8.979	8.046	8.026	8.904		1.100	0.950	0.932	0.930		0.950		1.023	0.952		0.811	0.832	Cu 29
30 Zn	9.659	8.637	8.614	9.570	9.656	1.196	1.043	1.021	1.012		1.034		1.107	1.044		0.884	0.906	Zn 30
31 Ga	10.368	9.250	9.223	10.263	10.365	1.300	1.142	1.117	1.098		1.125		1.197	1.134		0.957	0.984	Ga 31
32 Ge	11.104	9.885	9.854	10.980	11.099	1.420	1.248	1.218	1.188		1.218		1.294	1.249		1.036	1.068	Ge 32
33 As	11.868	10.542	10.506	11.724	11.862	1.530	1.359	1.325	1.282		1.317		1.388	1.360		1.120	1.155	As 33
34 Se	12.658	11.220	11.179	12.494	12.650	1.653	1.475	1.436	1.379		1.419		1.490	1.477		1.204	1.244	Se 34
35 Br	13.474	11.922	11.876	13.289	13.467	1.794	1.596	1.550	1.480		1.526		1.596	1.600		1.293	1.339	Br 35
36 Kr	14.322	12.648	12.596	14.110	14.312	1.920	1.756	1.675	1.586		1.636		1.706	1.728		1.387	1.439	Kr 36
37 Rb	15.201	13.393	13.333	14.959	15.183	2.067	1.866	1.806	1.694	1.692	1.752		1.826	1.863		1.482	1.542	Rb 37
38 Sr	16.105	14.163	14.095	15.833	16.082	2.216	2.007	1.940	1.806	1.804	1.871		1.947	2.003		1.582	1.649	Sr 38
39 Y	17.037	14.956	14.880	16.735	17.013	2.369	2.145	2.079	1.922	1.920	1.995		2.072	2.150		1.685	1.761	Y 39
40 Zr	17.998	15.772	15.688	17.665	17.967	2.547	2.307	2.223	2.042	2.040	2.124	2.219	2.201	2.302		1.792	1.876	Zr 40
41 Nb	18.986	16.612	16.518	18.619	18.949	2.698	2.465	2.371	2.166	2.163	2.257	2.367	2.334	2.461		1.902	1.996	Nb 41
42 Mo	20.002	17.476	17.371	19.605	19.962	2.866	2.625	2.520	2.293	2.289	2.394	2.518	2.473	2.623		2.015	2.120	Mo 42
43 Tc	21.054	18.364	18.248	20.615	21.002	3.054	2.795	2.677	2.424	2.421	2.536	2.674	2.616	2.790		2.131	2.248	Tc 43
44 Ru	22.118	19.276	19.147	21.653	22.070	3.236	2.966	2.837	2.558	2.554	2.683	2.835	2.763	2.964		2.252	2.382	Ru 44
45 Rh	23.224	20.213	20.070	22.720	23.169	3.419	3.146	3.003	2.696	2.692	2.834	3.001	2.915	3.143		2.376	2.519	Rh 45
46 Pd	24.350	21.174	21.017	23.815	24.295	3.617	3.330	3.173	2.838	2.833	2.990	3.171	3.072	3.328		2.503	2.660	Pd 46
47 Ag	25.514	22.159	21.987	24.938	25.452	3.806	3.524	3.351	2.984	2.978	3.150	3.347	3.234	3.519	3.749	2.633	2.806	Ag 47
48 Cd	26.711	23.170	22.980	26.091	26.639	4.019	3.737	3.537	3.133	3.126	3.316	3.528	3.401	3.716	3.751	2.767	2.956	Cd 48
49 In	27.940	24.206	23.998	27.271	27.856	4.237	3.938	3.730	3.286	3.279	3.487	3.713	3.572	3.920	4.160	2.904	3.112	In 49
50 Sn	29.200	25.267	25.040	28.481	29.104	4.465	4.156	3.929	3.443	3.435	3.662	3.904	3.750	4.130	4.376	3.044	3.272	Sn 50

TABLE 2 (continued)

Absorption Edge and X-Ray Energies

	K - SERIES					L - SERIES											M - SERIES					
Z Element	Kab	Kα1	Kα2	Kβ1	Kβ2	LIab	LIIab	LIIIab	Lα1	Lα2	Lβ1	Lβ2	Lβ3	Lγ1	Lγ3	Lℓ	Lη	MIVab	MVab	Mα1	Mα2	Mβ1	
51 Sb	30.486	26.357	26.109	29.721	30.388	4.698	4.381	4.132	3.604	3.595	3.843	4.100	3.932	4.347	4.598	3.188	3.436	0.537	0.528				51 Sb
52 Te	31.809	27.471	27.200	30.990	31.698	4.939	4.612	4.341	3.769	3.758	4.029	4.301	4.120	4.570	4.828	3.335	3.605	0.582	0.572				52 Te
53 I	33.164	28.610	28.315	32.289	33.036	5.188	4.852	4.557	3.937	3.925	4.220	4.507	4.313	4.800	5.065	3.484	3.780	0.631	0.620				53 I
54 Xe	34.579	29.802	29.485	33.619	34.408	5.452	5.100	4.781	4.109	4.096	4.417	4.718	4.512	5.036	5.305	3.637	3.958	0.685	0.672				54 Xe
55 Cs	35.959	30.970	30.623	34.981	35.815	5.720	5.358	5.011	4.286	4.272	4.619	4.935	4.716	5.279	5.552	3.794	4.141	0.740	0.726				55 Cs
56 Ba	37.410	32.191	31.815	36.372	37.251	5.995	5.624	5.247	4.465	4.450	4.827	5.156	4.926	5.530	5.808	3.953	4.330	0.796	0.781				56 Ba
57 La	38.931	33.440	33.033	37.795	38.723	6.267	5.891	5.483	4.650	4.633	5.041	5.383	5.143	5.788	6.073	4.124	4.524	0.851	0.833			0.854	57 La
58 Ce	40.449	34.717	34.276	39.251	40.226	6.549	6.165	5.724	4.839	4.822	5.261	5.612	5.364	6.051	6.340	4.287	4.731	0.902	0.884			0.902	58 Ce
59 Pr	41.998	36.023	35.548	40.741	41.767	6.846	6.443	5.968	5.033	5.013	5.488	5.849	5.591	6.321	6.615	4.452	4.935	0.953	0.932			0.949	59 Pr
60 Nd	43.571	37.359	36.845	42.264	43.327	7.126	6.722	6.208	5.229	5.207	5.721	6.088	5.828	6.601	6.900	4.632	5.145	1.001	0.977			0.996	60 Nd
61 Pm	45.207	38.649	38.160	43.818	44.929	7.448	7.018	6.466	5.432	5.407	5.960	6.338	6.070	6.891	7.188	4.813	5.363	1.052	1.027			1.047	61 Pm
62 Sm	46.846	40.124	39.523	45.405	46.566	7.737	7.312	6.717	5.635	5.607	6.205	6.586	6.317	7.177	7.485	4.994	5.588	1.106	1.078			1.100	62 Sm
63 Eu	48.515	41.529	40.877	47.030	48.248	8.069	7.624	6.983	5.845	5.816	6.455	6.842	6.570	7.479	7.795	5.176	5.816	1.161	1.131			1.153	63 Eu
64 Gd	50.229	42.983	42.280	48.688	49.952	8.376	7.931	7.243	6.056	6.024	6.712	7.102	6.830	7.784	8.104	5.361	6.049	1.217	1.185			1.209	64 Gd
65 Tb	51.998	44.470	43.737	50.374	51.715	8.708	8.252	7.515	6.272	6.237	6.977	7.365	7.095	8.100	8.422	5.546	6.283	1.274	1.240			1.266	65 Tb
66 Dy	53.789	45.985	45.200	52.110	53.500	9.083	8.621	7.850	6.494	6.457	7.246	7.634	7.369	8.417	8.752	5.742	6.533	1.332	1.295			1.325	66 Dy
67 Ho	55.615	47.528	46.692	53.868	55.315	9.395	8.919	8.071	6.719	6.679	7.524	7.910	7.650	8.746	9.086	5.942	6.787	1.392	1.351			1.383	67 Ho
68 Er	57.483	49.100	48.213	55.672	57.204	9.776	9.263	8.364	6.947	6.904	7.809	8.188	7.938	9.087	9.429	6.152	7.057	1.453	1.409			1.443	68 Er
69 Tm	59.335	50.735	49.766	57.506	59.085	10.116	9.618	8.648	7.179	7.132	8.100	8.467	8.229	9.424	9.778	6.341	7.308	1.515	1.468			1.503	69 Tm
70 Yb	61.303	52.363	51.329	59.356	60.974	10.486	9.978	8.943	7.414	7.366	8.400	8.757	8.535	9.778	10.141	6.544	7.579	1.577	1.528			1.567	70 Yb
71 Lu	63.304	54.065	52.960	61.272	62.956	10.867	10.345	9.241	7.654	7.604	8.708	9.047	8.845	10.142	10.509	6.752	7.856	1.641	1.591			1.631	71 Lu
72 Hf	65.313	55.762	54.584	63.222	64.969	11.264	10.739	9.561	7.898	7.843	9.021	9.346	9.162	10.514	10.889	6.958	8.138	1.716	1.662			1.697	72 Hf
73 Ta	67.400	57.529	56.273	65.212	67.001	11.680	11.139	9.881	8.145	8.086	9.342	9.650	9.486	10.893	11.276	7.172	8.427	1.793	1.735			1.765	73 Ta
74 W	69.508	59.315	57.979	67.233	69.089	12.098	11.542	10.204	8.396	8.334	9.671	9.960	9.817	11.284	11.672	7.386	8.723	1.871	1.809	1.775	1.773	1.835	74 W
75 Re	71.662	61.138	59.716	69.298	71.219	12.522	11.955	10.531	8.651	8.565	10.008	10.274	10.158	11.683	12.080	7.602	9.026	1.950	1.883	1.843	1.840	1.906	75 Re
76 Os	73.860	62.999	61.485	71.401	73.390	12.965	12.383	10.869	8.910	8.840	10.354	10.597	10.509	12.093	12.498	7.821	9.335	2.031	1.960	1.911	1.907	1.978	76 Os
77 Ir	76.097	64.895	63.286	73.548	75.606	13.424	12.824	11.215	9.174	9.098	10.706	10.919	10.866	12.510	12.922	8.040	9.649	2.116	2.041	1.980	1.975	2.053	77 Ir
78 Pt	78.379	66.832	65.123	75.735	77.864	13.892	13.273	11.564	9.441	9.360	11.069	11.249	11.233	12.940	13.359	8.267	9.973	2.202	2.122	2.050	2.046	2.127	78 Pt
79 Au	80.713	68.804	66.990	77.971	80.172	14.353	13.733	11.918	9.712	9.626	11.440	11.583	11.608	13.379	13.807	8.493	10.307	2.291	2.206	2.123	2.118	2.204	79 Au
80 Hg	83.106	70.819	68.895	80.240	82.530	14.846	14.209	12.284	9.987	9.896	11.821	11.922	11.993	13.828	14.262	8.720	10.649	2.385	2.295	2.196	2.191	2.282	80 Hg
81 Tl	85.517	72.871	70.832	82.562	84.933	15.344	14.698	12.657	10.267	10.171	12.211	12.270	12.388	14.289	14.734	8.952	10.992	2.485	2.389	2.270	2.265	2.362	81 Tl
82 Pb	88.001	74.969	72.804	84.922	87.343	15.860	15.198	13.035	10.550	10.448	12.612	12.621	12.791	14.762	15.215	9.183	11.347	2.586	2.484	2.345	2.339	2.442	82 Pb
83 Bi	90.521	77.108	74.814	87.335	89.833	16.385	15.708	13.418	10.837	10.729	13.021	12.978	13.208	15.245	15.708	9.419	11.710	2.687	2.579	2.422	2.416	2.525	83 Bi
84 Po	93.112	79.292	76.863	89.809	92.386	16.935	16.244	13.817	11.129	11.014	13.445	13.338	13.635	15.741	16.213	9.662	12.080	2.798	2.683	2.501	2.494	2.609	84 Po
85 At	95.740	81.513	78.943	92.319	94.976	17.490	16.784	14.215	11.425	11.303	13.874	13.703	14.065	16.249	16.720	9.902	12.458	2.909	2.787	2.580	2.573	2.695	85 At
86 Rn	98.418	83.782	81.064	94.877	97.616	18.058	17.337	14.618	11.725	11.596	14.313	14.073	14.509	16.768	17.248	10.141	12.844	3.022	2.892	2.661	2.653	2.782	86 Rn
87 Fr	101.147	86.104	83.230	97.483	100.305	18.638	17.904	15.028	12.029	11.893	14.768	14.448	14.973	17.300	17.800	10.381	13.248	3.136	3.000	2.743	2.735	2.870	87 Fr
88 Ra	103.927	88.469	85.433	100.136	103.048	19.233	19.078	15.442	12.338	12.194	15.233	14.839	15.442	17.845	18.354	10.620	13.661	3.253	3.109	2.826	2.817	2.961	88 Ra
89 Ac	106.759	90.884	87.674	102.846	105.838	19.842	19.078	15.865	12.650	12.499	15.710	15.230	15.929	18.405	18.925	10.863	14.080	3.371	3.219	2.910	2.901	3.052	89 Ac
90 Th	109.630	93.350	89.953	105.592	108.671	20.470	19.692	16.300	12.967	12.807	16.199	15.621	16.423	18.979	19.503	11.117	14.507	3.491	3.332	2.996	2.986	3.145	90 Th
91 Pa	112.581	95.869	92.286	108.409	111.575	21.102	20.311	16.731	13.288	13.120	16.699	16.022	16.927	19.565	20.094	11.364	14.944	3.609	3.442	3.082	3.072	3.239	91 Pa
92 U	115.591	98.439	94.665	111.289	114.549	21.756	20.947	17.167	13.612	13.437	17.217	16.425	17.452	20.164	20.709	11.616	15.397	3.728	3.552	3.170	3.159	3.336	92 U
93 Np	118.619	101.085	97.095	114.181	117.533	22.417	21.596	17.614	13.942	13.757	17.740	16.837	17.986	20.781	21.336	11.887	15.874	3.850	3.664	3.260	3.249	3.434	93 Np
94 Pu	121.720	103.750	99.536	117.146	120.592	23.095	22.263	18.053	14.276	14.082	18.278	17.254	18.537	21.414	21.979	12.122	16.330	3.973	3.778	3.352	3.340	3.533	94 Pu
95 Am	124.990	106.483	102.044	120.291	123.706	23.805	22.946	18.507	14.618	14.413	18.852	17.677	19.106	22.066	22.644	12.382	16.821	4.094	3.889	3.446	3.434	3.633	95 Am
96 Cm	128.251	109.279	104.598	123.409	126.890	24.529	23.653	18.972	14.959	14.745	19.426	18.104	19.687	22.731	23.319	12.636	17.317	4.227	4.013				96 Cm
97 Bk	131.570	112.131	107.198	126.587	130.150	25.266	24.372	19.439	15.307	15.082	20.015	18.537	20.283	23.414	24.014	12.894	17.827	4.357	4.132				97 Bk
98 Cf	134.952	115.047	109.848	129.834	133.500	26.015	25.104	19.905	15.659	15.423	20.622	18.976	20.897	24.195	24.726	13.150	18.349	4.482	4.246				98 Cf
99 Es	138.408	118.025	112.547	133.146	136.900	26.788	25.861	20.383	16.015	15.768	21.246	19.419	21.526	24.834	25.956	13.408	18.886	4.615	4.368				99 Es
100 Fm	141.943	121.071	115.300	136.529	140.370	27.584	26.643	20.872	16.375	16.115	21.886	19.867	22.170	25.572	26.200	13.666	19.437	4.757	4.497				100 Fm

where M X-rays are encountered is the analysis of sulfur when lead is present in the same sample. In this case, the Kα X-ray for sulfur appears at 2.31 keV and Kβ at 2.46 keV, hence the lead M X-rays at 2.39 keV overlap the sulfur K X-rays.

C. X-Ray Intensities and Fluorescence Yields

It has already been mentioned that an X-ray's energy is the single most important parameter in making qualitative identifications of those elements present in a sample. However, there are cases where qualitative identification of an element's presence may also depend on quantitative information in the spectra. This information is obtained from the measured net peak intensity of the observed X-rays. For example, if any two of the elements Ti, V, Cr, or Mn are present in a sample, their Kβ X-ray energies are almost identical to the Kα X-rays of the next higher Z adjacent element, i.e., V, Cr, Mn, and Fe. Thus, the quantitative evaluation of a small amount of iron in the presence of ten times as much manganese requires an accurate knowledge of the Kβ-to-Kα ratio for manganese. Other examples where energy information alone is insufficient are the exact energy overlap of the Pb Lα and As Kα X-rays, and the misidentification of titanium and vanadium X-rays in a sample when, in fact, the X-rays observed are really the Lα and Lβ X-rays of barium that are being fluoresced. Clearly then, quantitative information, and occasionally even unequivocal qualitative identification, will depend on measured X-ray intensities.

The intensity of any given X-ray line or component is governed by two fundamental atomic parameters. These are:

1. The electron population of higher lying levels, which may fill a lower lying vacancy, and
2. The mode of deexcitation or readjustment the atom takes.

Once an electron vacancy is created in the K shell (or any other shell), the atom is in an excited state. No atomic transition can occur to deexcite the atom unless higher lying shells, L, M, ... etc., are populated. Thus, hydrogen and helium have no X-rays because their L shells are not populated. For all elements equal to and above lithium, $Z = 3$, a Kα X-ray can be observed. Since Kβ X-rays correspond to electron transitions from the M or N shell to the K shell, no Kβ X-ray will occur until Z becomes equal to or greater than 11 (sodium). In fact, the Kβ X-ray commonly known as Kβ1, which is actually made up of (Kβ1 + Kβ3 + Kβ5′ + Kβ5″), does not reach full strength until all of the M subshells are filled, i.e., at $Z = 29$, copper (see Table 1). Only beyond $Z = 30$ (zinc) does the Kβ2 component begin to be observed (i.e. transitions from the N to K shell). It does not reach full strength until $Z = 46$ (palladium), which corresponds to the complete filling of the NII through NV subshells, since Kβ2 = (Kβ2′ + Kβ2″ + Kβ4). An examination of Table 2 will show that no L-series X-rays occur until $Z \geq 20$ (calcium), while M-series X-rays occur for $Z \geq 57$ (lanthanum).

The second fundamental parameter governing X-ray line intensity is how the atom deexcites. Once a vacancy is formed, it is not always followed by the emission of an X-ray. Three processes may occur. The atom may deexcite through the emission of an X-ray; or the atom may deexcite through the emission of an Auger electron, sometimes called a radiationless transition; or a transfer of vacancies can occur from one subshell to a less tightly bound subshell within a single major shell by radiationless Coster-Kronig transitions.

If the atom does not always emit an X-ray, then the yield of X-rays must be less than the number of vacancies formed. This yield of X-rays is known as the fluorescence yield and is denoted by ω. The fluorescence yield is the probability that a vacancy in an atomic shell or subshell will be filled through a radiative transition, i.e., an X-ray. Since there are no subshells for the K shell the following equation may be written

$$\omega_k = I_k / n_k \quad (1)$$

where ω_k is the fluorescence yield of the K shell, I_k is the total number of characteristic K X-rays, and n_k is the number of primary K-shell vacancies per unit time. Clearly, if $\omega_k < 1$, what happens the remainder of the time? For the K shell, the atom deexcites through the emission of an Auger electron from the L, M, N ... shell. Thus, if A_k is the yield of Auger electrons, then

$$A_k + \omega_k = 1 \quad (2)$$

Sometimes Auger electron emission is described as a two-step process in which the atom first emits an X-ray, but before it leaves the atom of origin, transfers all of its energy in ejecting an L, M, or N ... electron. Actually, it is a one-step process in which the atom deexcites by transferring an electron vacancy from the K shell to a less tightly bound shell with ejection of one of its electrons. This shell may now contain two electron vacancies, leaving the atom in a doubly ionized state. If one of the vacancies is subsequently filled by radiative transition, the emitted X-ray will have a slightly higher energy than that same transition has when falling into a single vacancy. These X-rays are known as satellite lines and are only observed in high-resolution wavelength dispersive spectrometers. For lighter atoms, Auger electron emission is much more probable than X-ray emission. Figure 3 illustrates the behavior of the K-fluorescence yield with increasing Z. For calcium (Z = 20) the K-fluorescence yield is about 14%; for germanium (Z = 32) it becomes greater than 50%, and above Z = 60, ω_k is greater than 90%.[20]

If a primary electron vacancy is created in one of the L or M subshells, then fluorescence and Auger yields are no longer easy to calculate. This arises because of the possibility of radiationless Coster-Kronig transitions. These are atomic transitions in which the vacancy remains in the same major shell where it was created, but an electron is emitted from this or any other less tightly bound shell. Thus, an LI vacancy may move to the LII or LIII subshell via a Coster-Kronig transition and electron ejection may occur in either of these two subshells or any M, N, ... subshell. Similarly, an LII

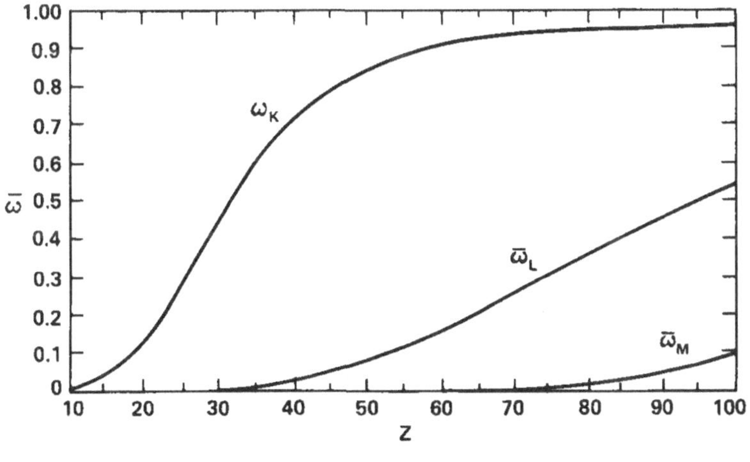

FIGURE 3. Fluorescence yields as a function of atomic number. Only mean yields can be given for the L and M shells. Exact yields depend on how the initial electron vacancy distribution is formed and filled.

vacancy may move to LIII via a Coster-Kronig transition, but an LIII vacancy can be filled only by X-ray emission or by Auger electron emission. Hence, the Auger yields of LI and LII subshells are modified by Coster-Kronig probabilities, and coupled with LIII Auger yields, determine the mean fluorescence yields, $\overline{\omega_L}$, for the entire L-shell. Figure 3 also shows the mean fluorescence yields, $\overline{\omega_L}$ and $\overline{\omega_M}$, for the L and M shells, respectively.

Table 3 lists the predicted K X-ray intensities relative to $K\alpha 1 = 100$ intensity units for all elements from $Z = 10$ to 100. These intensities are based on the calculations of Scofield.[21] Measured[22] $K\beta$ to $K\alpha$ ratios appear to exceed the theoretical predictions by 20% at low Z, and by about 7% for very high Z. Intensity ratios involving neighboring subshells, i.e, $K\alpha 2$ to $K\alpha 1$ or $L\alpha 2$ to $L\alpha 1$, appear to agree well with theory. The last column gives the predicted intensity of $L\alpha 1$ to $K\alpha 1$. Table 4 lists the predicted L X-ray intensities relative to $L\alpha 1 = 100$ for all elements from $Z = 28$ to 100. Again these data are taken from Scofield's[21] calculations. In most cases of FEA, the three $L\beta$ components will not be resolved and should be added together. For L X-rays which follow certain nuclear processes (nculear internal conversion and L-electron capture), the $L\beta$ components cannot be added because the vacancies formed may be restricted to a particular subshell. In general, the theoretical intensities listed in Tables 3 and 4 are best taken as guides to the expected intensities. Self-absorption of X-rays within a sample will also affect the observed intensity. Hence, the usual practice is to carefully prepare standards, which are as similar to the unknown as possible and which contain known amounts of trace elements, and then use the resulting measured X-ray intensities as reference values. A more detailed discussion of fluorescence yields, Auger electron yields, and Coster-Kronig transition probabilities may be found in the excellent review article by Bambynek et al.[23]

The three factors just discussed describe how the atom will deexcite, but do not address how the vacancy is formed. Vacancy creation can affect the average fluorescence yield for any shell and thus influence subsequent X-ray intensities. Two different experiments may not give rise to the same fluorescence yield, ω; and therefore, lead to different X-ray intensities if the ionization methods used create different primary vacancy distributions. Thus, ω is not a fundamental property of the atom, but depends on the atomic subshell configurations and on parameters peculiar to the experiment.

III. FLUORESCENT EXCITATION ANALYSIS

In the phenomenological description of FEA presented in Section I, the two most important ways in which radiation interacts with matter were introduced. The first was photoelectric absorption of the radiation and the second was radiation scattering. This section will discuss these two interaction modes in more detail because they are essential to an understanding of FEA. A third important way in which radiation can interact with matter is called pair production, but it leads only indirectly to the production of X-rays and will not be discussed. In the discussion that follows, excitation by only photon sources with single electron vacancy creation will be assumed. The various methods of atomic excitation are discussed in Chapter 3.

A. Radiation Absorption and Attenuation

Consider a perfectly collimated and monochromatic radiation beam of intensity I_o incident on matter having thickness x (cm) and density ρ (g/cm^3). The emergent beam will have an intensity I, which is always less than I_o, since the radiation beam undergoes absorption or attenuation in passing through matter. The rate of change of the beam intensity is given by

$$dI/I = -\mu\, dx \tag{3}$$

TABLE 3

Theoretical K X Emission Rates (Intensities)

$K\alpha1$ = 100 Intensity units

Z	$K\alpha2$	$K\beta1$[a]	$K\beta2$[b]	$K\delta1,2$	Z	$L\alpha1$[c]
10	50.4				10	
12	50.5				12	
14	50.5				14	
16	50.6	7.0			16	
18	50.7	12.3			18	
20	50.7	16.0			20	
22	50.9	17.1			22	
24	50.9	17.1			24	
26	51.1	18.1			26	
28	51.2	18.4			28	1.05
30	51.3	18.8			30	1.37
32	51.6	19.7	0.32		32	1.66
34	51.7	20.6	1.12		34	1.92
36	51.9	21.6	1.88		36	2.15
38	52.1	22.4	2.57		38	2.38
40	52.3	23.3	3.10		40	2.59
42	52.5	24.1	3.45		42	2.78
44	52.7	24.9	3.76		44	2.95
46	53.0	25.5	4.11		46	3.11
48	53.3	26.2	4.48		48	3.27
50	53.4	26.7	4.83	0.10	50	3.42
52	53.8	27.3	5.18	0.38	52	3.55
54	54.1	27.8	5.57	0.67	54	3.68
56	54.3	28.3	5.91	0.85	56	3.81
58	54.7	28.8	6.11	0.86	58	3.91
60	54.9	29.2	6.30	0.87	60	4.06
62	55.2	29.7	6.45	0.87	62	4.17
64	55.6	30.1	6.61	0.87	64	4.28
66	56.0	30.5	6.75	0.87	66	4.38
68	56.4	30.9	6.89	0.86	68	4.48
70	56.8	31.4	7.00	0.86	70	4.58
72	57.2	31.7	7.18	0.95	72	4.67
74	57.6	32.1	7.35	1.07	74	4.76
76	58.1	32.5	7.52	1.16	76	4.84
78	58.5	32.8	7.68	1.26	78	4.93
80	59.0	33.1	7.88	1.37	80	5.01
82	59.6	33.5	8.07	1.49	82	5.09
84	60.0	33.8	8.24	1.60	84	5.17
86	60.6	34.1	8.44	1.72	86	5.25
88	61.1	34.4	8.61	1.84	88	5.33
90	61.9	34.7	8.81	1.96	98	5.41
92	62.6	35.1	8.98	2.07	92	5.49
94	63.0	35.4	9.17	2.20	94	5.57
96	63.8	35.7	9.36	2.32	96	5.65
98	64.5	36.1	9.53	2.45	98	5.72
100	65.4	36.4	9.73	2.58	100	5.80

[a] $K\beta1 = K\beta1 + K\beta3 + K\beta5' + K\beta5''$.

[b] $K\beta2 = K\beta2' + K\beta2'' + KBA$.

[c] $L\alpha1$ relative to $K\alpha1$ = 100 intensity units.

TABLE 4

Theoretical L X-ray Emission Rates (Intensities)

$L\alpha 1$ = 100 Intensity units

Z	$L\alpha 2$	$L\beta 1$[a]	$L\beta 2$[b]	$L\beta 3$[c]	$L\gamma 1$[d]	$L\gamma 3$[e]	$L\ell$	$L\eta$	Z
28	11.0	57.0		35.6			7.5	3.7	28
30	11.3	56.5		29.8			6.1	2.9	30
32	11.3	56.0		26.8			5.4	2.5	32
34	11.3	57.0		25.1			5.0	2.2	34
36	11.3	57.5	0.6	24.2	0.3	2.3	4.7	2.1	36
38	11.3	57.5	0.6	23.5	0.3	3.0	4.5	2.0	38
40	11.3	57.5	2.0	23.1	1.0	3.4	4.3	1.9	40
42	11.3	58.0	4.5	22.8	2.3	3.6	4.2	1.8	42
44	11.3	58.5	7.1	22.6	3.8	3.8	4.2	1.8	44
46	11.3	58.5	9.7	22.6	5.2	4.0	4.2	1.8	46
48	11.3	59.0	12.0	22.5	6.3	4.1	4.1	1.7	48
50	11.3	59.0	14.0	22.4	7.4	4.5	4.1	1.7	50
52	11.3	59.5	15.9	22.4	8.6	5.0	4.2	1.7	52
54	11.3	60.0	17.5	22.4	9.5	5.4	4.2	1.7	54
56	11.3	60.0	18.7	22.4	10.1	5.8	4.3	1.7	56
58	11.3	60.5	19.2	22.4	10.4	5.9	4.4	1.7	58
60	11.3	60.5	19.4	22.3	10.5	6.0	4.4	1.7	60
62	11.3	61.0	19.7	22.3	10.8	6.0	4.5	1.6	62
64	11.3	61.5	19.9	22.4	11.0	6.1	4.6	1.6	64
66	11.3	61.5	20.1	22.5	11.2	6.1	4.7	1.7	66
68	11.3	62.0	20.2	22.5	11.3	6.2	4.8	1.7	68
70	11.3	62.5	20.3	22.6	11.6	6.2	4.9	1.7	70
72	11.3	63.1	20.9	22.7	12.1	6.4	5.0	1.7	72
74	11.3	63.6	21.7	22.8	12.6	6.6	5.2	1.7	74
76	11.3	64.1	22.8	22.8	13.4	6.7	5.3	1.7	76
78	11.3	64.1	23.8	22.8	14.0	6.9	5.5	1.7	78
80	11.4	64.6	24.8	22.9	14.8	7.0	5.7	1.7	80
82	11.4	65.1	25.9	23.0	15.6	7.2	5.9	1.8	82
84	11.4	65.6	26.7	23.1	16.3	7.4	6.1	1.8	84
86	11.4	66.1	27.5	23.2	17.1	7.6	6.3	1.8	86
88	11.4	66.6	28.5	23.3	17.8	7.8	6.5	1.9	88
90	11.4	67.6	29.4	23.3	18.4	8.0	6.7	1.9	90
92	11.4	68.1	30.0	23.4	19.1	8.2	6.9	1.9	92
94	11.4	68.6	30.8	23.4	19.8	8.3	7.2	2.0	94
96	11.4	69.1	31.6	23.4	20.5	8.5	7.4	2.0	96
98	11.4	69.6	32.3	23.5	21.2	8.7	7.6	2.0	98
100	11.4	70.1	32.9	23.6	21.7	8.9	7.8	2.0	100

[a] $L\beta 1 = L\beta 1 + L\beta 17$.
[b] $L\beta 2 = L\beta 2 + L\beta 5 + L\beta 6 + L\beta 7 + L\beta 15 + Lu$.
[c] $L\beta 3 = L\beta 3 + L\beta 4 + L\beta 9 + L\beta 10$.
[d] $L\gamma 1 = L\gamma 1 + L\gamma 5 + L\gamma 8 + Lv$.
[e] $L\gamma 3 = L\gamma 2 + L\gamma 3 + L\gamma 4' + L\gamma 4 + L\gamma 11$.

where μ is known as the linear absorption coefficient and has units cm^{-1}. The negative sign indicates that the intensity always decreases as it passes through matter. If μ is independent of x, integration of Equation 3 gives

$$I = I_0 \exp(-\mu x) \qquad (4)$$

It is more useful to rewrite Equation 4 in terms of the mass absorption coefficient, μ_m, where

$$\mu_m(\text{cm}^2/\text{gm}) = \mu(\text{cm}^{-1}) / \rho\ (\text{g/cm}^3) \quad (5)$$

Thus Equation 5 becomes

$$I = I_0 \exp(-\mu_m \rho x) \quad (6)$$

The quantity μ_m is related to the probability that radiation will interact with matter in some manner. This probability or cross-section may be thought of as a measure of the area of the atom (in a mathematical sense) for the radiation interaction mode in question. The fundamental unit of atomic cross-section is the barn which is 10^{-24} cm^2. Tabulations of the mass absorption coefficient for all elements, are often given in terms of barns/atom, that is,

$$\mu_m(\text{cm}^2/\text{g}) = \sigma(\text{barns/atom}) \frac{N\ (\text{atoms/mol})}{A\ (\text{g/mol})} \quad (7)$$

where N is Avogardro's number (6.02×10^{23} atoms/mol) and A is the atomic weight. Clearly, the conversion from σ to μ_m is through a multiplicative constant which varies for each element and is included in tabulations of σ.

The linear absorption coefficient, and therefore the mass absorption coefficient, give a measure of the total absorption of the radiation which passes through any material, regardless of the interaction mode. Since there are three primary interaction modes, μ_m can be written as

$$\mu_m = \mu/\rho = (\tau + \sigma_s + \pi)/\rho \quad (8)$$

where τ represents the photoelectric cross-section, σ_s represents the total scattering cross section, and π represents the pair-production cross section. Since the latter is zero until the radiation energy exceeds 1 MeV, only the photoelectric and scattering cross sections are responsible for absorption or attenuation of radiation below 1 MeV.

B. Photoelectric Absorption

The classical picture of an atom surrounded by shells and subshells of electrons suggests that the electrons are bound by discrete energies. Quantum mechanically this is in fact the case. The innermost electrons are bound most tightly while the outer valence electrons are only loosely bound. The more loosely bound an electron, the lower the incoming radiation energy need be in order to eject it. As the exciting radiation energy is increased, it suddenly becomes sufficient to eject electrons from the next inner subshell or shell. At this point the absorption cross-section abruptly increases, giving rise to increased absorption. If the radiation energy is increased still further, the absorption coefficient decreases approximately as E^{-3} until a more tightly bound shell is encountered. This abrupt increase in the cross-section is referred to as an absorption edge and such an edge is associated with each atomic shell or subshell. Figure 4 shows a log-log plot of the photoelectric absorption cross-section in barns per atom for lead. The N-, O-, and P-shell absorption edges occur well below 1 keV. Above 1 keV, the absorption edge for the loosely bound MV subshell electrons occurs first at 2.48 keV. Then come less dramatic increases for the other M subshells. Just above 13 keV, the LIII absorption edge occurs, followed by the LII and LI edges. Finally, the K-shell absorption edge occurs at 88-keV. For any given exciting radiation energy, electrons will be ejected from all shells as long as the incident energy is greater than

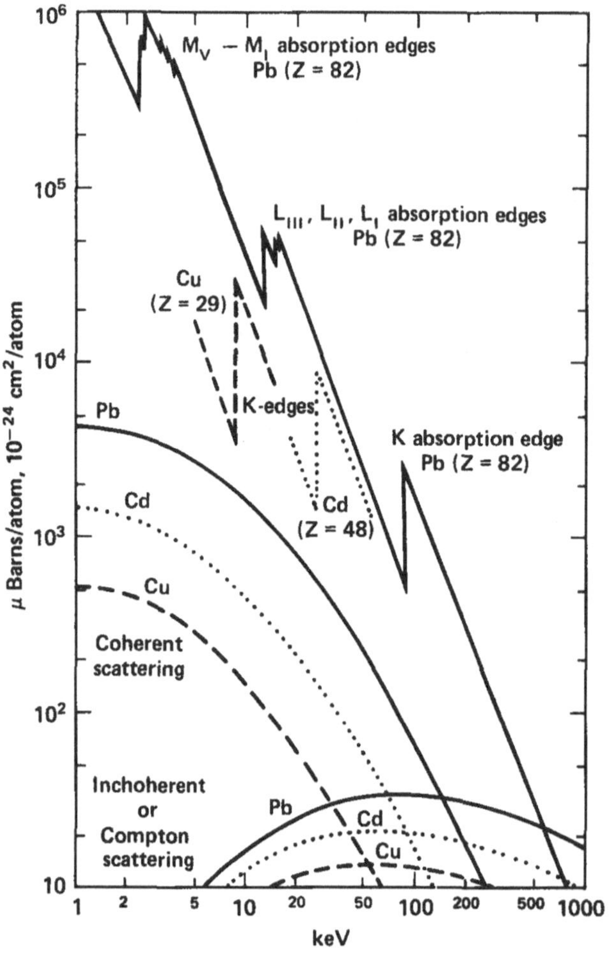

FIGURE 4. The atomic absorption and scattering cross-sections in barns (10^{-24}cm²) per atom as a function of energy for lead (Z = 82), cadmium (Z = 48), and copper (Z = 29). The L-absorption edge positions for copper and cadmium are not shown. Barns/atom can be converted into cm²/g via the multiplicative constants of 2.91×10^{-3} for cadmium, and 9.48×10^{-3} for copper.

the shell binding energy. Thus, for 100-keV radiation incident on lead, approximately 80% of the electrons ejected will be from the K shell, 14% from the three L subshells, about 5% from the five M subshells, and the remainder from the N, O, and P subshells.

As the atomic number decreases, the K-shell (and all other) electrons become less tightly bound. Correspondingly, the K-absorption edges decrease in energy as well. Figure 4 also shows the position and magnitude in barns/atom of the K absorption edges of cadmium (Z = 48) and copper (Z = 29). Note that the K edge for copper occurs at a lower energy than the L edges of lead. A source of 22-keV exciting radiation will eject electrons from lead creating predominantly L-shell vacancies, which give rise to the L series of X-rays; while this same radiation will eject primarily K-electrons (and some L-electrons) from copper leading to K X-rays (L X-rays). Tabulations of photon cross-sections for energies from 1-keV to 100-MeV and for all elements from hydrogen (Z = 1) to fermium (Z = 100) are published by Storm and Israel.[24] The K, LI, LII, LIII, MIV, and MV absorption edges for all elements from Z = 1 to 100 are listed in Table 2.

C. Coherent Radiation Scattering

When a beam of photon radiation passes through matter, each interaction must be visualized as taking place in the presence of a single atom. When a single photon interacts with an atom's group of electrons, it can be scattered with or without energy loss. Scattering without energy loss is known as Rayleigh or coherent scattering. In this type of scattering, the outgoing radiation has a definite phase relationship with the incident radiation. It is better understood as if the incident wave of radiation sets up an oscillation of the entire atomic electron system; and then the atom reirradiates or scatters the incident radiation in any direction. For a given radiation energy, the more electrons an atom has, the greater the coherent scattering cross-section. On the other hand, the greater the incident radiation energy, the less probable coherent scattering is for a given atomic number. The energy and atomic number dependence of the coherent scattering cross-section are shown by the three curves for copper, cadmium, and lead (Z = 29, 48, and 82) in Figure 4.

The importance of coherent scattering in FEA is that the exciting radiation usually appears in the resultant spectrum of the sample being excited, regardless of the geometrical relationship between the exciting source and the detector. This follows because coherent scattering does not change the energy of the incident radiation, but can change its direction.

D. Incoherent Radiation Scattering

Radiation scattering in which there is energy loss is known as Compton scattering or incoherent radiation scattering. In this atomic process the single photon can be visualized as interacting in a particle-like manner with a single atomic electron. The incident photon suffers a change in direction, an atomic electron is given kinetic energy which may be sufficient to eject it from the atom, and the outgoing radiation has less energy. Conservation of energy and momentum applied to this billiard-ball-like collision process leads to the following relationship between the incident photon energy, E_o, and the scattered photon energy, E'.

$$E' = E_0 \left[1 + \frac{E_0}{511} (1 - \cos\theta) \right]^{-1} \qquad (9)$$

E_o and E' are expressed in keV, 511 is the rest mass energy (m_oc^2) of the electron in keV, and θ is the angle through which the photon is scattered. The scattering angle can take on values anywhere between 0 and 180°. Comptom scattered photons lose energy and may be recorded in the spectrum of interest at E', which is always less than or equal to E_o. The energy shift, ΔE, will depend on the scattering angle and varies from 0 keV for $\theta = 0°$ to a maximum at 180° (a backscattered photon). The order of magnitude of this energy shift is shown in Figure 5 for four different photon energies. Note that for Mn Kα radiation at 5.9 keV, the maximum energy shift is less than 140 eV; i.e., the Compton or incoherent scattered radiation will not be visible in the spectrum as a separate peak in energy dispersive systems. Rather, it simply broadens the unshifted, coherently scattered peak. On the other hand, for 22-keV exciting radiation, the scattered radiation peak can easily be resolved from the coherent or unmodified peak. Figure 5 shows that for scattering angles from 50 to 180°, the energy shift for 22-keV radiation varies from 0.3 keV to 1.8 keV. As the incident energy increases, the energy shift for a given scattering angle increases and the Compton or incoherent peak is easily separated from the coherent peak, which diminishes in intensity. The incoherent peak will also become broader than the coherent peak, a result of a range of scattering angles being accepted by the detector. The intensity of the incoherent peak can be used[25,26] to obtain information on the effective Z of the sample and on matrix absorption effects.

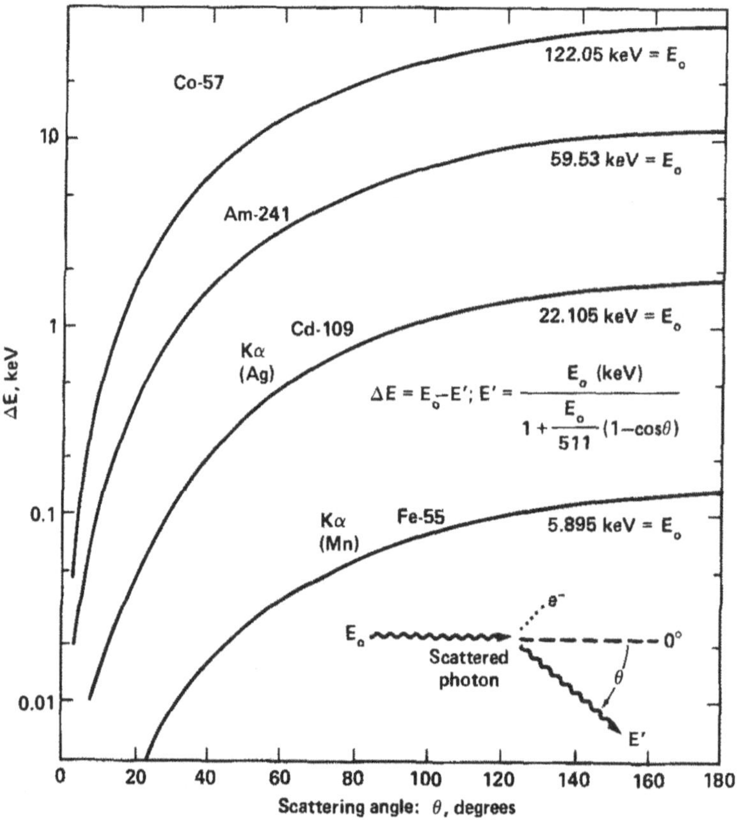

FIGURE 5. Energy shift in keV of Compton scattered photons as a function of scattering angle, θ, for four particular photon energies.

Radiation scattering is not equally probable for all angles. It depends on both θ and E. Figure 6 illustrates the probability, P, that a quantum will be scattered through an angle θ relative to $\theta = 0°$. For low energies, $E < 22$ keV, scattering is more probable in the forward and backward directions, and about half as probable, relatively, near 90°. As energy increases, the probability for scattering it at angles greater than 90° decreases. This figure demonstrates why in most energy dispersive spectroscopy systems, angular differences near 90° are chosen between the exciter-sample and the sample-detector collimation axes. This choice minimizes the amount of scattered radiation reaching the detector from the sample; hence maximizes the fluoresced to scattered radiation detected. It is interesting to note from Figure 6 that although small angle, forward scattering is likely from collimator walls, the associated energy shift shown in Figure 5 is very small to negligible ($\Delta E = 0.2$ keV at 5° for $E_o = 122$ keV, or 0.02 keV at 5° for 22 keV).

E. Polarized Radiation

The incoherent or Compton scattered peak in a spectrum is often very intense, especially for thick or massive samples. In searching for very weak trace element concentrations, which are at the threshold of detectability, events associated with the intense incoherent scatter peak can nearly mask the X-ray of interest. One way to reduce the intensity of the scatter peak is to use incident radiation which has been polarized. Radiation which emerges from an X-ray tube or from a radioactive source is not polarized. That is, the direction of the electric vector, which is part of the mathematical description of electromagnetic radiation, is randomly oriented in space. However,

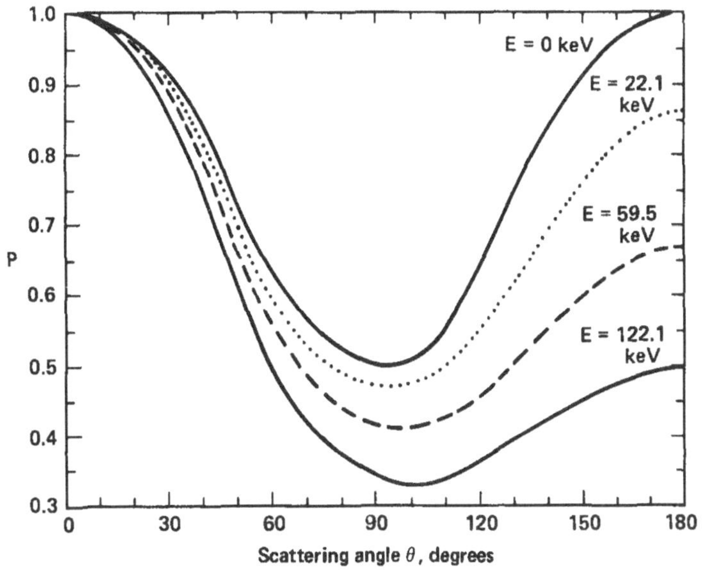

FIGURE 6. Relative probability of Compton scattering per unit solid angle as a function of scattering angle, θ, for three particular photon energies.

when unpolarized radiation is scattered at 90°, the scattered photon will have its electric vector perpendicular to the plane of scattering. Thus, a beam of photons scattered at 90° is said to be polarized. The best polarizing materials will have low Z (low absorption cross-section), but high density (more electrons/cm^3) to enhance scattering. Typical materials which can be used are lucite, carbon, beryllium, and aluminum. Kaufman and Camp[27] have shown that boron carbide is one of the best polarizing materials for 60-keV radiation.

Figure 7 shows collimated photons of energy E_o emitted from an X-ray source along the Z-axis and incident on a polarizing material. The initial beam is scattered at all angles, but if by collimation a portion of the incoherently scattered radiation is selected along the X-axis, the resulting beam of energy E′ will have its electric vector *E* perpendicular to the XZ-plane. The modified energy, E′, can be calculated from Equation 9 using $\theta = 90°$. By definition, polarized radiation cannot scatter in the direction of its electric vector, i.e., into its own plane of polarization. Thus, if the polarized beam of Figure 7 is incident on a sample and if the detector-sample collimation axis is chosen as the Y-axis, i.e., normal to the XZ-scattering plane, then very little of the polarized radiation can be scattered into the detector. Meanwhile, part of the polarized radiation will excite or fluoresce X-rays from trace elements within the sample. Since the fluoresced X-rays are emitted in all directions, part of the events will be recorded by the detector. The resultant incoherent scatter peak will be considerably reduced in intensity relative to what it would have been if the sample had been irradiated directly by the X-ray source. However, its intensity does not go to zero because the beam is only incompletely polarized (partially controlled by the collimation geometry) and because of multiple small angle scattering in the polarizer and sample (if $E_o = 60$ keV, then $E' = 54$ keV for $\theta = 90°$ while $E' = 56$ keV for $\theta_1 = 45°$ and $\theta_2 = 45°$). When ordinary fluorescent excitation analysis is carried out on nonthin samples, the incoherent or Compton scatter peak is often quite intense. In these cases, polarized fluorescence analysis can be quite effective since the incoherent scattering peak is considerably reduced in intensity.

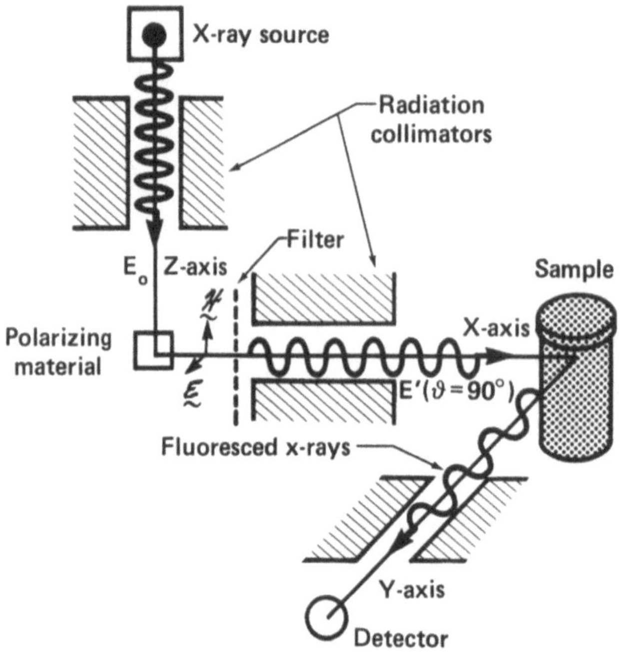

FIGURE 7. Geometrical arrangement for creating polarized exciting radiation. The Z-axis photon beam is unpolarized but after scattering and selection, a beam along the X-axis is polarized. The filter is used to harden or reduce low energy scattered photons. Compton scattered intensity will be a minimum in the Y-axis direction, allowing the most sensitivity to fluoresced X-rays from the sample.

F. Summary

The analytical technique of fluorescent excitation analysis can be used for the rapid, nondestructive qualitative and quantitative analysis of a wide range of sample types and geometries. The inorganic elements contained in a sample can be stimulated to emit their own characteristic X-rays by excitation with suitable radiation. The exciting radiation is scattered or absorbed by the sample, producing an emission spectrum.

Such an emission spectrum recorded by a lithium drifted silicon detector is shown in Figure 8. The individual events recorded by the detector and processed by appropriate electronics (see Chapter 2) vary from zero energy up to a maximum, i.e., the exciting radiation energy. This variation in energy is shown as the horizontal scale or abscissae in Figure 8. The number of events or counts per channel number (energy interval) recorded by the multichannel analyzer is shown as the vertical scale or ordinate in Figure 8. By choosing a semilogarithmic vertical scale, a greater dynamic range in counts per channel can be displayed.

The sample, which gave the spectrum shown in Figure 8, was a 1.5-cm diameter by 2.5-cm tall plastic vial containing water with 12.5 mg of iodine dissolved per gram of solution. The exciting radiation was the 59.53-keV gamma ray from a radioactive source of Am-241. From Table 2, the binding energy of a K electron in iodine is shown as 33.164 keV. Since the exciting radiation energy is greater than the K-electron binding energy, a K-electron vacancy can be produced in iodine if an exciting photon is photoelectrically absorbed by the iodine atom.

From Figure 3, we see that the fluorescence yield for iodine ($Z = 53$) is 0.87, that is, a K-shell vacancy will lead to a K-series X-ray 87% of the time. From Table 2, the $K\alpha 1$ and $K\alpha 2$ X-rays are found to have energies 28.610 and 28.315 keV, respectively, which are too close to be resolved into separate peaks by the detector used. The $K\beta 1$

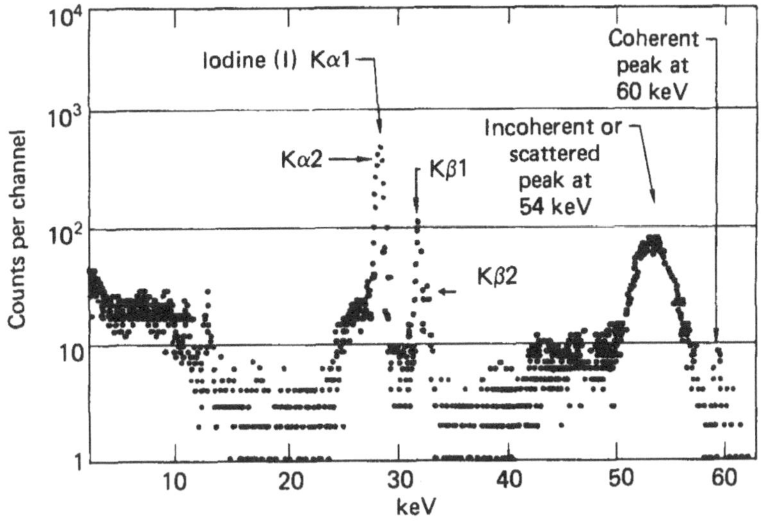

FIGURE 8. A fluorescence excitation analysis (FEA) spectrum of a sample containing 12.5 mgI/g solution. The excitation source is a 60-keV gamma ray from radioactive Am-241. The gamma rays are collimated to strike the sample directly. The entire spectrum was obtained in 5 sec. Note that the ordinate scale is semilogarithmic, not linear.

and Kβ2 have energies 32.289 and 33.036 keV, respectively, which are resolved, but not clearly evident on the small scale of Figure 8.

The intensities of these four X-rays can be interpolated from the relative intensities listed in Table 3 for even Z. For Z = 53 we have, relative to a Kα1 intensity of 100 units, approximate intensities of 54 for Kα1, 27.6 for Kβ1, and 5.4 for Kβ2 of iodine. The Kδ X-rays are too weak to be observed. Any L-series X-rays of iodine produced, i.e., Lα, Lβ, or Lγ, are all below 4.0 keV and are self-absorbed within the relatively massive sample. The suggestive peaks seen at 10.5 and 12.5 keV are Lα and Lβ X-rays of lead from the collimator materials used in the experiment.

Some evidence for the coherently scattered 60 keV exciting radiation is just observable in Figure 8. Had more data been accumulated, this sharp peak would have been readily apparent. The broad distribution centered about 54 keV corresponds to incoherently or Compton scattered 59.5-keV radiation. From Figure 5 it can be seen that 59.5-keV exciting radiation scattered through an angle of 90° will lose 5.5 keV; thus, the detector sees primarily Compton scattered 54-keV radiation. The broad width of this scattered radiation peak is evidence that the experimental geometry allowed the detector to accept a wide range of scattering angles.

This same plastic vial sample was arranged according to the experimental geometry illustrated in Figure 7. The 59.5-keV exciting radiation, E_o, was collimated along the Z-axis and scattered off of boron carbide. By using lead collimators to select a scattering angle of 90°, the scattered radiation becomes polarized in the ZX-plane, and has a mean energy of 54 keV. This polarized radiation strikes the sample, and if the detector is placed orthogonal to the ZX-plane, i.e., along the Y-axis, the X-ray emission spectrum from the sample may be observed with a minimum contribution from incoherently scattered radiation. The result is shown as the top spectrum (A) in Figure 9. Data were accumulated in this spectrum until the unresolved Kα1 to Kα2 X-ray peaks of iodine had the same number of total counts as obtained in the spectrum of Figure 8. This spectrum shows much fewer counts at both low energies (<10 keV) and in the region of the incoherent peak, now centered at about 48 keV, relative to the intensity of Kα X-rays of iodine.

If the detector is moved from the Y-axis into the ZX-plane and located along the Z-

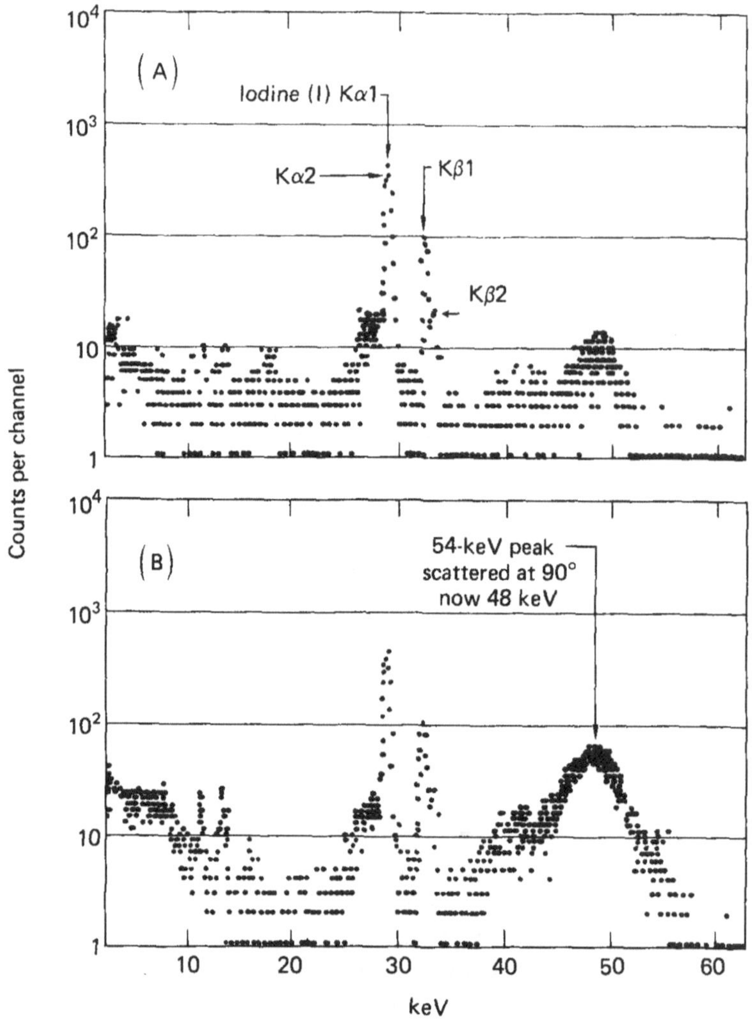

FIGURE 9. Two 1000-sec FEA spectra of a sample containing 12.5 mgI/g solution. In spectrum (A) the 60-keV exciting radiation is polarized exactly as shown by the geometrical arrangement in Figure 1.7. In spectrum (B) the detector is moved from the Y-axis direction to the Z-axis direction, defeating the polarization condition. Note that the incoherent peak in (B) is significantly larger than in (A). Introduction of the polarizer has broadened and shifted the scattered radiation peak down in energy relative to its width and position as seen in Figure 8.

axis, the emission spectrum recorded becomes that shown in Figure 9(B). The reduction in the incoherent peak intensity gained by polarization of the exciting radiation has been forfeited. Note that the 48-keV incoherent peak distribution here is even wider than that of the 54-keV incoherent peak distribution of Figure 8, a result of a second 90° scattering. Note that Lα and Lβ X-rays arising from the lead collimators are easily observed in (B), and there is even evidence of Lγ1 from lead at 14.76 keV.

Data accumulation using the polarization geometry required 1000 sec while that of Figure 8 required only 5 sec, a factor of 200. Clearly, for concentration levels in the mg/g level polarized exciting radiation is not needed. However, had the iodine concentration been 1.25 μg/g solution, that is, about 1 ppm or less, the iodine Kα1,2, peaks would not have been observable under the conditions of Figure 8, but would have under the conditions of Figure 9(A). Therefore, polarized exciting radiation offers an improvement in analytical sensitivity for trace element concentrations near the threshold of detectability in relatively massive samples.

PERIODIC TABLE OF THE ELEMENTS

PERIODIC TABLE OF THE ELEMENTS

KEY TO CHART

Atomic Number → **50** +2 ← Oxidation States
Symbol → Sn +4
Atomic Weight → 118.69
18 18 4 ← Electron Configuration

1a	2a	3b	4b	5b	6b	7b	8			1b	2b	3a	4a	5a	6a	7a	0	Orbit
1 H +1 -1 1.0079 1																	**2** He 0 4.00260 2	K
3 Li +1 6.94 2-1	**4** Be +2 9.01218 2-2											**5** B +3 10.81 2-3	**6** C +2 +4 -4 12.011 2-4	**7** N +1 +2 +3 +4 +5 -1 -2 -3 14.0067 2-5	**8** O -2 15.9994 2-6	**9** F -1 18.998403 2-7	**10** Ne 0 20.17₉ 2-8	K-L
11 Na +1 22.98977 2-8-1	**12** Mg +2 24.305 2-8-2	Transition Elements					Group 8			Transition Elements		**13** Al +3 26.98154 2-8-3	**14** Si +2 +4 -4 28.0855 2-8-4	**15** P +3 +5 -3 30.97376 2-8-5	**16** S +4 +6 -2 32.06 2-8-6	**17** Cl +1 +5 +7 -1 35.453 2-8-7	**18** Ar 0 39.948 2-8-8	K-L-M
19 K +1 39.0983 -8-8-1	**20** Ca +2 40.08 -8-8-2	**21** Sc +3 44.9559 -8-9-2	**22** Ti +2 +3 +4 47.90 -8-10-2	**23** V +2 +3 +4 +5 50.9415 -8-11-2	**24** Cr +2 +3 +6 51.996 -8-13-1	**25** Mn +2 +3 +4 +7 54.9380 -8-13-2	**26** Fe +2 +3 55.847 -8-14-2	**27** Co +2 +3 58.9332 -8-15-2	**28** Ni +2 +3 58.71 -8-16-2	**29** Cu +1 +2 63.546 -8-18-1	**30** Zn +2 65.38 -8-18-2	**31** Ga +3 69.735 -8-18-3	**32** Ge +2 +4 72.59 -8-18-4	**33** As +3 +5 -3 74.9216 -8-18-5	**34** Se +4 +6 -2 78.96 -8-18-6	**35** Br +1 +5 -1 79.904 -8-18-7	**36** Kr 0 83.80 -8-18-8	-L-M-N
37 Rb +1 85.467₈ -18-8-1	**38** Sr +2 87.62 -18-8-2	**39** Y +3 88.9059 -18-9-2	**40** Zr +4 91.22 -18-10-2	**41** Nb +3 +5 92.9064 -18-12-1	**42** Mo +6 95.94 -18-13-1	**43** Tc +4 +6 +7 98.9062 -18-13-2	**44** Ru +3 101.07 -18-15-1	**45** Rh +3 102.9055 -18-16-1	**46** Pd +2 +4 106.4 -18-18-0	**47** Ag +1 107.868 -18-18-1	**48** Cd +2 112.41 -18-18-2	**49** In +3 114.82 -18-18-3	**50** Sn +2 +4 118.69 -18-18-4	**51** Sb +3 +5 -3 121.75 -18-18-5	**52** Te +4 +6 -2 127.60 -18-18-6	**53** I +1 +5 +7 -1 126.9045 -18-18-7	**54** Xe 0 131.30 -18-18-8	-M-N-O
55 Cs +1 132.9054 -18-8-1	**56** Ba +2 137.33 -18-8-2	**57*** La +3 138.9055 -18-9-2	**72** Hf +4 178.49 -32-10-2	**73** Ta +5 180.947₉ -32-11-2	**74** W +6 183.85 -32-12-2	**75** Re +4 +6 +7 186.2 -32-13-2	**76** Os +3 +4 190.2 -32-14-2	**77** Ir +3 +4 192.22 -32-15-2	**78** Pt +2 +4 195.09 -32-16-2	**79** Au +1 +3 196.9665 -32-18-1	**80** Hg +1 +2 200.59 -32-18-2	**81** Tl +1 +3 204.37 -32-18-3	**82** Pb +2 +4 207.2 -32-18-4	**83** Bi +3 +5 208.9808 -32-18-5	**84** Po +2 +4 (209) -32-18-6	**85** At (210) -32-18-7	**86** Rn 0 (222) -32-18-8	-N-O-P
87 Fr +1 (223) -18-8-1	**88** Ra +2 226.0254 -18-8-2	**89**** Ac +3 (227) -18-9-2	**104** — +4 (261) -32-10-2	**105** — (262) -32-11-2	106 (263) 32-12-2													O P Q

															Orbit
*Lanthanides	**58** Ce +3 +4 140.12 -20-8-2	**59** Pr +3 140.9077 -21-8-2	**60** Nd +3 144.24 -22-8-2	**61** Pm +3 (145) -23-8-2	**62** Sm +2 +3 150.4 -24-8-2	**63** Eu +2 +3 151.96 -25-8-2	**64** Gd +3 157.25 -25-9-2	**65** Tb +3 158.9254 -27-8-2	**66** Dy +3 162.50 -28-8-2	**67** Ho +3 164.9304 -29-8-2	**68** Er +3 167.26 -30-8-2	**69** Tm +3 168.9342 -31-8-2	**70** Yb +2 +3 173.04 -32-8-2	**71** Lu +3 174.967 ± 0.003 -32-9-2	N O P
**Actinides	**90** Th +4 232.0381 -18-10-2	**91** Pa +5 +4 231.0359 -20-9-2	**92** U +3 +4 +5 +6 238.029 -21-9-2	**93** Np +3 +4 +5 +6 237.0482 -22-9-2	**94** Pu +3 +4 +5 +6 (244) -24-8-2	**95** Am +3 +4 +5 +6 (243) -25-8-2	**96** Cm +3 (247) -25-9-2	**97** Bk +3 +4 (247) -27-8-2	**98** Cf +3 (251) -28-8-2	**99** Es +3 (254) -29-8-2	**100** Fm +3 (257) -30-8-2	**101** Md +2 +3 (258) -31-8-2	**102** No +2 +3 (259) -32-8-2	**103** Lr +3 (260) -32-9-2	O P Q

Numbers in parentheses are mass numbers of most stable isotope of that element

Appendix 2
THE ELEMENTS AND THEIR SYMBOLS

Name	Symbol	Name	Symbol
Actinum	Ac	Molybdenum	Mo
Aluminum	Al	Neodymium	Nd
Americium	Am	Neon	Ne
Antimony	Sb	Neptunium	Np
Argon	Ar	Nickel	Ni
Arsenic (gray)	As	Niobium (Columbium)	Nb
Astatine	At		
Barium	Ba	Nitrogen	N
Berkelium	Bk	Nobelium	No
Beryllium	Be	Osmium	Os
Bismuth	Bi	Oxygen	O
Boron	B	Palladium	Pd
Bromine	Br	Phosphorus	P
Cadmium	Cd	Platinum	Pt
Calcium	Ca	Plutonium	Pu
Californium	Cf	Polonium	Po
Carbon	C	Potassium	K
Cerium	Ce	Praeseodymium	Pr
Cesium	Cs	Promethium	Pm
Chlorine	Cl	Protactinium	Pa
Chromium	Cr	Radium	Ra
Cobalt	Co	Radon	Rn
Copper	Cu	Rhenium	Re
Curium	Cm	Rhodium	Rh
Dysprosium	Dy	Rubidium	Rb
Einsteinium	Es	Ruthenium	Ru
Erbium	Er	Samarium	Sm
Europium	Eu	Scandium	Sc
Fermium	Fm	Selenium	Se
Fluorine	F	Silicon	Si
Francium	Fr	Silver	Ag
Gadolinium	Gd	Sodium	Na
Gallium	Ga	Strontium	Sr
Germanium	Ge	Sulfur	S
Gold	Au	Tantalum	Ta
Hafnium	Hf	Technetium	Tc
Helium	He	Tellurium	Te
Holmium	Ho	Terbium	Tb
Hydrogen	H	Thallium	Tl
Indium	In	Thorium	Th
Iodine	I	Thulium	Tm
Iridium	Ir	Tin	Sn
Iron	Fe	Titanium	Ti
Krypton	Kr	Tungsten	W
Lanthanum	La	Uranium	U
Lawrencium	Lr	Vanadium	V
Lead	Pb	Wolfram (see Tungsten)	
Lithium	Li		
Lutetium	Lu	Xenon	Xe
Magnesium	Mg	Ytterbium	Yb
Manganese	Mn	Yttrium	Y
Mendelevium	Md	Zinc	Zn
Mercury	Hg	Zirconium	Zr

REFERENCES

1. Roentgen, W. C., On a new kind of rays, Sitzungsber, *Phys. Med. Ges.*, Wurzburg, pg. 137, 1895.
2. Moseley, H. G. J., High frequency spectra of the elements, *Philos. Mag.*, 26, 1024, 1913; 27, 703, 1914.
3. Bohr, N., *Philos. Mag.*, 26, 426, 1913.
4. Hadding, A., Mineral analysis by x-ray spectroscopic methods, *Z. Inorg. Chem.*, 122, 195, 1922.
5. Bowman, H. R., Hyde, E. K., Thompson, S. G., and Jared, R. C., *Science*, 71, 562, 1966.
6. Bonner, N. A., Bazan, F., and Camp, D. C., Trace element analysis using x-ray fluorescence, *Chem. Instrum.*, 6, 1, 1975.
7. Dzubay, T. G., Ed., *X-Ray Fluorescence Analysis of Environmental Samples*, Ann Arbor Science, Ann Arbor, 1977, 310.
8. Ondov, J. M., Ragaini, R. C., and Biermann, A. H., Elemental emissions from a coal fired power plant — comparison of a Venturi wet scrubber system with a cold-side electrostatic precipitator, Lawrence Livermore Laboratory Report, UCRL-80110, *Environ. Sci. Tech.*, in press, 1978.
9. Cooper, J. A., Wheeler, B. D., and Bartell, D. M., Rapid multielemental analysis of Portland cement — raw mix and raw materials, *Cem. Technol.*, 7, 68, 1976.
10. Barrett, C. S., Leyden, D. E., Newkirk, J. B., and Ruud, C. O., Eds., Advances in x-Ray Analysis 21, Plenum Press, New York, 1978.
11. Langheinrich, A. P., Forster, J. W., and Linn, T. A., Jr., Current status and potential of solid state detectors in industrial and field applications, in *Nuclear Techniques in the Basic Metal Industries*, IAEA-SM-159/21, 429, 1973.
12. Frankel, R., Detection of art forgeries by x-ray fluorescence spectroscopy, *Isot. Radiat. Technol.*, 8, 1970.
13. Manson, V. F., Quantitative elemental analysis of art objects by energy dispersive x-ray fluorescence spectroscopy, *Appl. Spectrosc.*, 27, 309, 1973.
14. Artzy, M. and Asaro, F., Origin of Tell El-Yahudiyah Ware, Lawrence Berkeley Laboratory, LBL-4399, June, 1977. To be published in the *J. Am. Inst. Archaeol.*, 1978.
15. Artzy, M., Lawrence Berkeley Laboratory Report LBL-5017, April 1976. To be published in the Proc. Conference on Applications of Physical Science to Medieval Ceramics, UCLA, UCB, and LBL, March 18-22, 1976.
16. Cooper, J. A., The criminal leaves his card. *Ind. Res.*, 19 (Special), 22, 1977.
17. Hubbard, N. J. and Keith, J. E., Reevaluation of the apollo orbital x-ray fluorescence data, *Proc. Lunar Science Conference*, Vol. 8, Pergamon Press, New York, 909, 1977.
18. Clark, B. C., Baird, A. K., Rose, H. J., Toulmin, P., Keil, K., Castro, A. J., Kelliher, W. C., Rowe, C. O., and Evans, P. H., Inorganic analyses of martian samples at the Viking landing sites, *Science*, 194, 1283, 1976.
19. Lederer, C. M. and Shirley, V. S., *Table of Isotopes*, 7th ed., John Wiley & Sons, New York, 1978.
20. Bertin, E. P., *Principles & Practices of X-Ray Spectrometric Analysis*, 2nd ed., Plenum Press, New York, 1975, 980.
21. Scofield, J. H., *Phys. Rev.*, 179, 9, 1969.
22. Rao, P. V., Chen, M. H., and Crosemann, B., *Phys. Rev.*, A5, 997, 1972.
23. Bambynek, W., Cravemann, B., Fink, R. W., Freund, H. U., Mark, H., Swift, C. D., Price, R. E., and Rao, P. V., X-ray fluorescence yields, Auger, and Coster-Kronig transition probabilities, *Rev. Mod. Phys.*, 44, 716, 1972.
24. Storm, E. and Israel, H. I., Photon cross sections from 1 keV to 100 MeV for elements Z = 1 to Z = 100, *Nucl. Data*, 7, 565, 1970.
25. Reynolds, R. C., Jr., *Am. Mineral.*, 52, 1493, 1967.
26. Feather, C. E. and Willis, J. P., *X-Ray Spectrom.*, 4, 41, 1976.
27. Kaufman, L. and Camp, D. C., Polarized radiation for x-ray fluorescence analysis, in *Advances in X-Ray Analysis*, Vol. 18, Plenum Publishing, 1974, 247.

Chapter 2

INSTRUMENTATION FOR ENERGY DISPERSIVE X-RAY FLUORESCENCE

Joseph M. Jaklevic and Fred S. Goulding

TABLE OF CONTENTS

I. Introduction 29

II. Semiconductor Detectors 32
- A. General Design 32
- B. Detector Fabrication 32
- C. Signals, Noise, and Resolution 33
- D. Efficiency of Semiconductor Detectors 34
- E. Background 35
  - 1. X-Ray Escape Peak 35
  - 2. Continuum Background 36

III. Signal Processing Electronics 37
- A. Input Circuit — Preamplifier 37
  - 1. Choice of Field-Effect Transistor 37
  - 2. Charge-Sensitive Method 39
  - 3. Pulsed-Light Feedback 40
- B. Amplifier Requirements 41
  - 1. Gain, Linearity 41
  - 2. Pulse Shaping and Its Effect on Resolution 41
- C. System Considerations 44
- D. Pulsed-Excitation Technique 45

IV. Digital Processing 46

References 47

I. INTRODUCTION

X-ray fluorescence analysis depends on illuminating a sample with radiation of an energy and type adequate to produce inner-shell vacancies in the sample atoms. These vacancies then de-excite to produce X-rays whose energies are characteristics of the elements in the sample. The energy distribution of X-rays emitted from the sample can then be related to the concentrations of the elements present.

The energies of X-rays normally encountered in X-ray fluorescence analysis range from approximately 1 keV to 40 keV. For X-rays below 1 keV, the attenuation problems become so severe that accurate quantitative analysis becomes extremely difficult. For energies above 40 keV, excitation of characteristic K-shell fluorescence becomes improbable; consequently, L X-rays, which occur at low energies, are generally used for the heavier elements.

The viability of the X-ray fluorescence (XRF) method depends upon the availability of a detector system which can measure the energies and intensities of X-rays in this energy region. It must exhibit an energy resolution adequate to distinguish one element from the next one in the periodic table and, ideally, to reduce the number of ambiguous overlaps of spectral lines to a minimum. During most of the history of X-ray fluorescence analysis, spectra were measured using crystal spectrometers which depend on the Bragg reflection of X-rays from regular, periodic lattices. A beam of X-rays incident at an angle θ with respect to a regular crystal surface will be reflected if the wavelength λ satisfies the following condition:

$$\lambda = 2d \sin \theta \tag{1}$$

where d is the spacing between lattice planes in the crystal. By limiting the range of incident angles and counting the number of reflected X-rays as a function of angle of incidence, the spectral distribution of the radiation is determined. The wavelength and energy are related by the expression

$$\lambda(\text{Å}) = \frac{12.4}{E(\text{keV})} \tag{2}$$

where $1\text{Å} = 10^{-8}$ cm. By selecting the crystal with appropriate lattice spacing and by using slits to limit the angular divergence of the beam, it is possible to construct crystal spectrometers exhibiting an energy resolution of 10 eV or less in the energy range of interest. This is more than adequate to discriminate between adjacent elements in the periodic table and eliminate almost all ambiguities in the interpretation of spectra which arise from the overlap of complex K and L X-ray spectra.

This energy resolution is achieved at a considerable loss in intensity due to the restrictive geometry necessary to produce a narrow X-ray beam. Furthermore, since only one wavelength can be reflected at a given θ setting, the spectrometer is basically a single-channel instrument and must be sequentially scanned through the range of interest. The low efficiency of the spectrometer can be compensated by using a higher incident excitation flux. Multiple crystal spectrometers are also available for simultaneous analysis of more than one element. This technique, called wavelength-dispersive X-ray fluorescence analysis, has been in use for several decades.[1-3]

In the past 10 years, the development of semiconductor type radiation detectors for nuclear physics applications has resulted in the availability of another technique for XRF measurements.[4] These detectors convert the energy of incident photons into ionization in the semiconductor detector material. The ionization signal is closely proportional to the photo energy, a fact which makes these devices very convenient spectrometers for XRF analysis.

Although various detectors operating on induced ionization have been available for many years, it was not until modern semiconductor detectors were developed that the energy resolution became adequate for XRF analysis. Since their introduction in the early 1960s, semiconductor detector X-ray spectrometers have been continuously improved in energy resolution and counting-rate capabilities. The energy-dispersive X-ray method which has been developed around semiconductor detector spectrometers has supplanted the wavelength-dispersive method in a number of important applications. This chapter is devoted to the instrumentation associated with semiconductor detector X-ray spectrometers.

X-ray fluorescence is achieved by illuminating the sample with an appropriate source of electrons, photons, or charge particles. Fluorescence X-rays together with possible background radiation are emitted more or less isotropically from the sample. In energy-dispersive X-ray fluorescence, a semiconductor detector is placed near the sample in order to intercept and analyze a fraction of these X-rays.

A typical X-ray detector is a circular device with an active region 0.5 to 1.0 cm in diam and 3 to 5 mm thick. When X-rays with energies in the range of interest for XRF are incident on semiconductor materials such as silicon or germanium, they interact with the material principally by the K-shell photoelectric effect. This results in inner shell vacancies together with ejected photoelectrons whose energies are equal to that of the incident photon minus the K-shell binding energy. These vacancies deexcite either by emitting low-energy electrons or characteristic silicon or germanium X-rays which, in most cases, are then reabsorbed in the detector material. The energy of the electrons produced both in the photoelectric interaction and in the deexcitation process is absorbed in the crystal by repeated ionizing collisions with the atoms in the lattices which produce numerous free electrons and their associated positive holes. The number of electron-hole pairs produced is very large and there is a well defined average energy (ε) per pair. The number of free charges is therefore:

$$n = E/\epsilon \tag{3}$$

where E is the initial photon energy. The semiconductor materials used in detectors are of such quality that virtually all the free carriers are collected to produce a charge signal that is accurately proportional to the initial photon energy. This pulse is then electronically amplified and shaped so as to maximize the signal amplitude relative to any random noise sources. The pulse processing system is critical in achieving the energy resolution performance required for X-ray fluorescence spectrometry.

The electronic amplifier output consists of a random time sequence of pulses with varying amplitude reflecting the emission spectrum of the sample. The amplitude of each pulse represents the energy of an X-ray, and its frequency of occurrence is proportional to the intensity of that particular type of emission.

In order to readily interpret these data, individual pulse amplitudes are digitized and a histogram of pulse amplitudes is accumulated in a multichannel analyzer. The spectrum can then be analyzed to obtain energies and intensities of X-ray lines which can then be converted to the desired analytical information. Figure 1 is a typical X-ray

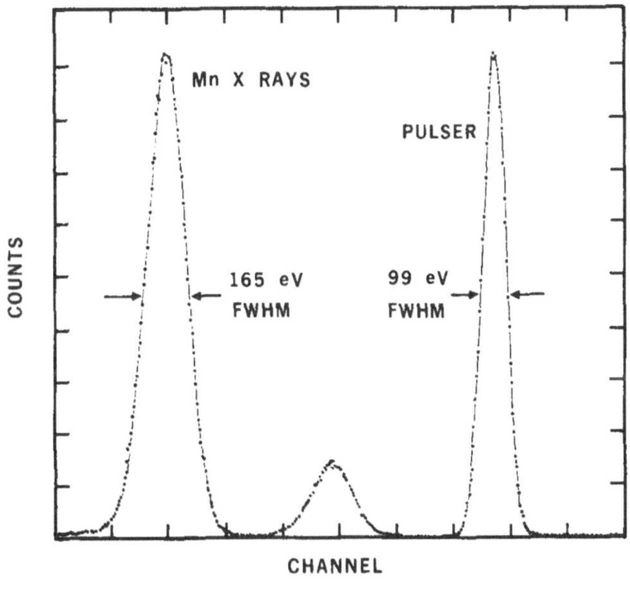

FIGURE 1. Typical x-ray spectrum showing manganese, Kα, and Kβ X-rays. The pulser peak is electronically generated and represents the noise contributing to total resolution.

spectrum showing the Kα and Kβ lines of manganese accumulated in a pulse-height analyzer.

It is important to note that the semiconductor detector simultaneously observes all X-rays incident on the detector. Therefore, a broad range of elements can simultaneously be measured with more or less uniform sensitivity. Qualitative analyses can easily be performed by a quick survey of the spectrum. Quantitative analyses require that an accurate sensitivity curve for the instrument is known.

II. SEMICONDUCTOR DETECTORS

A. General Design

The use of solid state semiconductor materials for radiation detectors results from their unique electrical properties. For any particular semiconductor there exists a temperature below which virtually no free charge carriers are present with which to sustain an electrical current when a voltage is applied. For silicon the critical temperature is well below room temperature while germanium requires lower temperatures (77°K — liquid nitrogen — is commonly used to cool both silicon and germanium spectrometers). Although the materials normally behave as insulators at this temperature, only a small amount of energy is necessary to release bound electrons that are then excited across the narrow band-gap into the conduction band where they can be collected by an applied voltage. Since the energy required to generate charge carriers is small, absorption of an energetic quantum of incident radiation results in a large number of free charges and a correspondingly large charge signal.

An important requirement of semiconductor materials for detector applications is the ability to collect all the free charge produced in photon interactions. Impurities and defects in crystals can create electrically active traps where free charges can be temporarily bound and effectively removed from the signal thus reducing its amplitude. Fortunately, it is possible to obtain materials of sufficient quality to perform suitably for detectors.

B. Detector Fabrication

Although many semiconductor materials have been explored for detector applications, all very high resolution spectrometers presently employ either silicon or germanium. In the case of Ge, it is possible to obtain material containing so few donors (i.e., atoms releasing electrons and therefore acting as fixed positive charges in the lattice) or acceptors (the inverse of donors) that relatively low applied voltages will deplete the material of free charge in order to form a sensitive region for radiation detection. Such high-purity germanium detectors are currently available although their use in X-ray applications is limited for reasons to be discussed. They are applied mainly to the detection of high-energy X-rays and γ-rays where the higher atomic number and correspondingly greater photoelectric cross sections of Ge contribute to increased absorption effciency.

Most semiconductor detector X-ray spectrometers used in X-ray fluorescence employ lithium-drifted Si detectors (Si(Li)). For that reason they will be described here in more detail than other types of devices; although, most of the general ideas regarding their properties will be applicable to others. Figure 2 shows a cross section of Si(Li) detector. Typical dimensions might be 2-cm total wafer diameter and 3- to 5-mm thickness. The active area is defined by the mesa structure in the middle and is typically 5 to 10 mm in diameter. The bulk of the detector volume constitutes the active region where events are detected. Since Si is not available with sufficiently low doping levels due to purification problems, it is necessary to compensate the acceptor sites in the detector volume. Starting with p-type material (i.e., mainly acceptor-doped), it is pos-

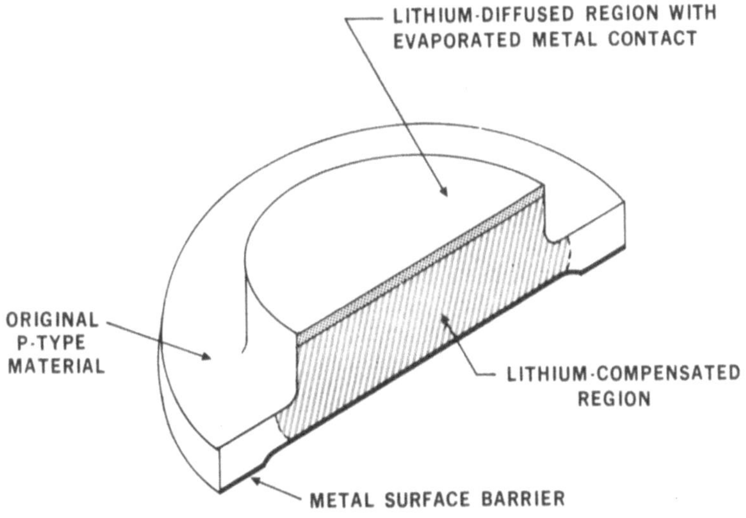

FIGURE 2. Cross section of a typical lithium-drifted silicon detector.

sible to drift lithium donor atoms into the volume under an applied field in such a way that they perfectly compensate the number of acceptors. The number of effective free charge carriers is thereby reduced almost to zero, and the application of relatively low voltages will deplete sufficient thickness of material (several mm) to serve as a detector.

To deplete the active volume and provide the electric field required to collect the radiation-induced free carriers (holes and electrons), it is necessary to apply a voltage of a few hundred volts across the detector bulk. For the detector to withstand this voltage without injecting charge from the contacts, opposite faces of the detector are heavily doped with p- and n-type impurities. The n-type contact is then biased at a positive voltage with respect to the p-type. Since the strongly doped contacts contain few minority carriers, virtually no current can be injected from the contacts into the detector volume. Such a detector is referred to as a p-i-n structure where i refers to intrinsic or carrier free region, and p and n refer to the two contacts.

C. Signals, Noise, and Resolution

Since the signal of interest consists of a pulse of charge collected across the detector, it is obvious that the accuracy of measurement of the signal is affected by spurious charges collected at the electrodes or otherwise introduced into the electronic system. One important consequence of this is the necessity to cool detectors to reduce the production of free carriers by thermal vibrations in the lattice. Since silicon and germanium have finite energy gaps across which charges can be excited, a number of free charge carriers are normally available to sustain a current with an applied voltage. By reducing the temperature, this leakage current is drastically reduced. For high-resolution X-ray spectrometers, detectors are normally operated in vacuum cryostats at temperatures approaching liquid nitrogen (77°K).

Although cooling reduces the detector leakage, it is unavoidable that some noise sources remain not only in the detector, but also arising from random fluctuations in currents in the later amplifying stages. Such noise sets a fundamental limit to the energy resolution which can be achieved. The electronic signal processing is designed to reduce this to a minimum; modern spectrometers are characterized by electronic contributions to resolution in the range of 60 to 100 eV full width at half maximum (FWHM).

The total energy resolution when observing X-rays is a combination of this electronic noise and a contribution due to statistical fluctuations in the ionization process in the detector. The total number of hole electron pairs produced by a particle of energy E has previously been defined by Equation 3. Since the formation of these charge pairs is a statistical process, there is a deviation in this number which can be expressed as

$$E_{FWHM} = 2.35\sqrt{E\epsilon F} \tag{4}$$

This equation is similar to that for a Poisson distributed variable with the exception of the factor F. This so-called Fano factor takes into account the fact that the interaction processes which produce free charge in the detector are not completely statistically independent. Measured values for F in Ge or Si range form 0.08 to 0.13. Combining this with a value of ε of 2.96 for germanium and 3.7 for silicon, the contribution due to charge-production statistics to the total resolution is approximately 100 eV for germanium and 120 eV for silicon at a photon energy of 5.9 keV.

The energy resolution of semiconductor spectrometers is the quadrature sum of electronic noise and statistical spread. The full width at half maximum ΔE, can be expressed as

$$\Delta E = \sqrt{E_N^2 + E_D^2} \tag{5}$$

where E_N is the FWHM contribution of electronic noise and E_D is due to charge-production statistics. Figure 3 is a plot of this relationship for a variety of cases.

D. Efficiency of Semiconductor Detectors

The total efficiency of a semiconductor detector X-ray spectrometer is determined by the product of geometric efficiency and intrinsic detector response. The geometric efficiency is a solid angle subtended by the detector at the sample to detector distance. For typical X-ray fluorescence spectrometer applications, detector to sample distances of 2 to 10 cm are employed, corresponding to geometric efficiencies in the range of 1 to 0.04%. The intrinsic efficiency of a detector is defined as the probability that an X-ray incident on the sensitive region of the detector will be absorbed and produce a signal. Figure 4 shows plots of the intrinsic efficiencies of typical Si(Li) and Ge detec-

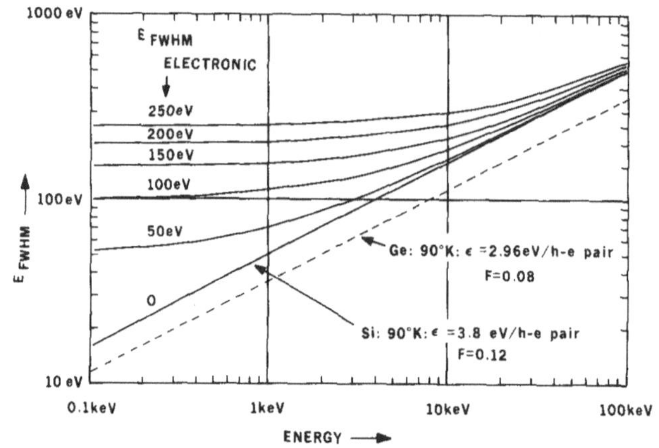

FIGURE 3. Resolution as a function of energy for several choices of parameters. Typical values of X-ray detectors might be 100 eV electronic noise with F = 0.12.

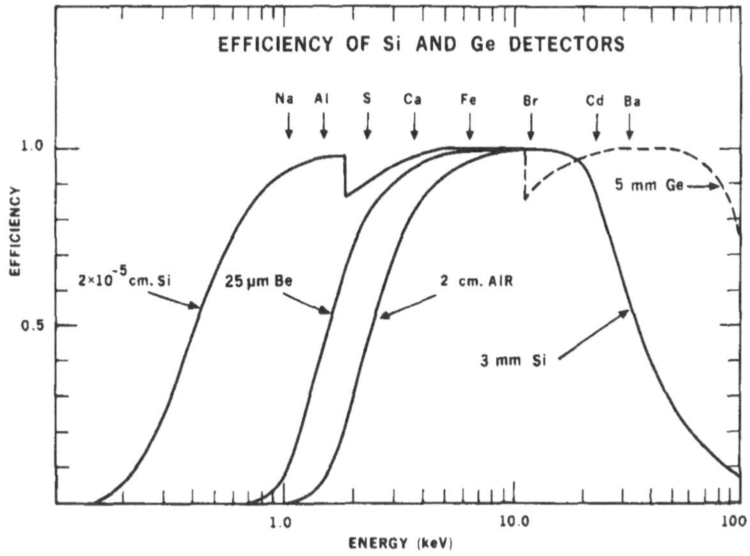

FIGURE 4. Efficiency vs. energy for a 3 mm-thick silicon detector and a 5 mm germanium detector. The low-energy limit to efficiency is determined by window absorption.

tors as a function of photon energy. The efficiency at low energies is limited by absorption in the entry window of the detector and in any materials (including air) between sample and detector. A typical entrance window on a Si(Li) detector consists of a 0.2 μm dead layer on the p^+ contact side. The cryostat entry window is typically a pure beryllium foil 25 μm thick. The remaining X-ray path may be either air, vacuum, or helium depending on the application. At the high-energy end of the curve, the efficiency is proportional to $(1-e^{-\mu\rho x})$ where $\mu\rho$ is the photoelectric cross section in cm^{-1} and x is the detector thickness.

E. Background

In the foregoing discussion, it has been assumed that the spectral response of a detector to monoenergetic X-rays is a single Gaussian-shaped peak with resolution determine by Equation 5. However, several mechanisms contribute to a more complex response.

1. X-Ray Escape Peak

The primary interaction of X-rays with the detector material results in the photoelectric ionization of an inner shell. The resulting vacancy often de-excites by emitting a characteristic X-ray of the material. The Kα energy for silicon is 1.74 keV while for germanium it is 9.8 keV. When photoelectric events occur deep within the detector, the characteristic detector material X-rays are reabsorbed and contribute to the full-energy signal. However, for events which occur near the surface, there is a significant probability that the characteristic radiations escapes and a corresponding amount of energy is lost from the signal. This results in escape peaks at energies 1.74 keV (Si) or 9.8 keV (Ge) below the full-energy peaks. The magnitude of these peaks is proportional to the fluorescence yield for the vacancies in the detector material and depends upon the absorption depth of the incident radiation. The peaks are most prominent when the incident photon energy is just greater than the absorption depth of the incident radiation. The peaks are most prominent with the incident photon energy is just greater than the absorption edge energy. Figure 5 is a spectrum of zirconium X-rays detected

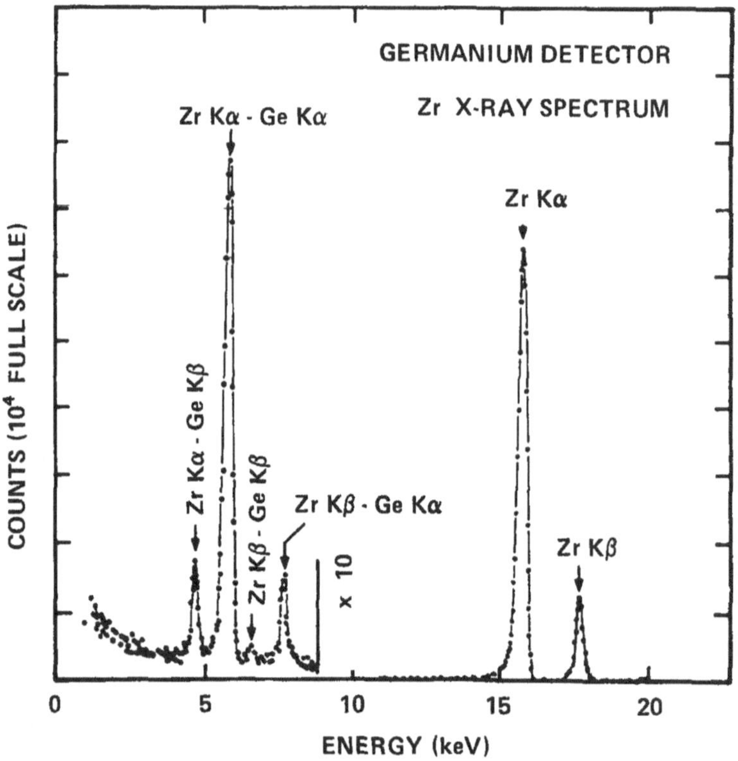

FIGURE 5. Spectrum of zirconium Kα and Kβ X-rays as measured with a germanium detector showing the system peaks due to radiation escape from the detector.

by a Ge detector showing the Kα and Kβ escape peaks. Since the fluorescence yield for the Ge K-shell is large (ω_K = 0.5), the magnitudes of the escape peaks are much larger than for silicon where ω_K = .06. The enhanced escape probability and its adverse effect on the interpretation of complex spectra complicates the interpretation of X-ray fluorescence spectra obtained with Ge detectors.

2. Continuum Background

All X-ray fluorescence spectra exhibit continuous background distributions which span the energy region of interest in XRF analysis. Since fluctuations in this background can hide real peaks, the identification and reduction of the effects which produce background are important objectives for improving analytical sensitivity. The complete process whereby the photon energy is converted into a charge signal involves a number of intermediate steps in which energetic electrons, holes, and photons are produced in the crystal. Any escape of radiation from the active volume, results in a background of partial amplitude pulses. In most applications, these fundamental background processes do not appear to seriously limit the use of the detectors.

There is one source of continuous background arising from detector artifacts which is potentially serious, but can be reduced by careful experimental design. This results from incomplete collection of the charge due to irregularities in the electrostatic field lines within the detectors.[5] Figure 6 shows a common type of detector geometry employed for X-ray spectrometers. Where the intrinsic material reached the surface of the device, it usually behaves as a lightly n-doped surface layer and the resulting internal electric field lines are distorted as shown. Since charge collection occurs along these field lines, some of the charge is collected via the surface layer. Once charge reaches

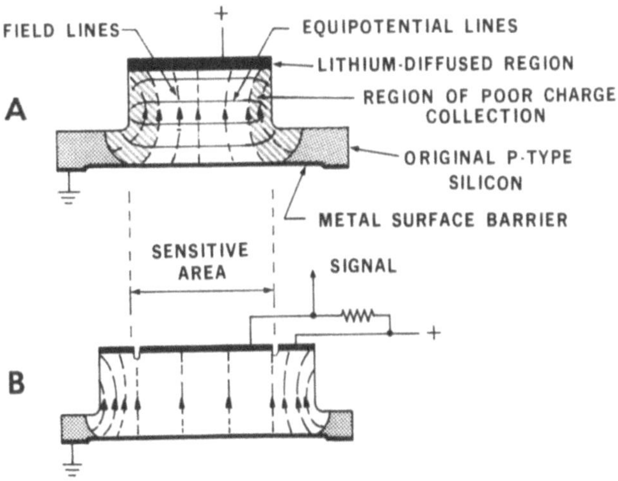

FIGURE 6. Field distribution with Si(Li) detectors. (A) Typical detector exhibiting field distortions due to surface channels. (B) Guard-ring detector showing a sensitive region defined by internal field lines.

the surface, it becomes trapped for sufficiently long times that it no longer contributes to the normal input signal. These events produce partial amplitude signals. In some detectors as many as 40% of the signals produced by 20 keV photons are degraded into a low-energy continuous background distribution. This effect can be reduced by the use of external collimation to restrict the effective area of the detector to a small central region[6] or by using internal electrostatic collimation in the form of a guard-ring detector as shown in Figure 6. Here a larger area device is fabricated, but the signal is extracted only from an electrically isolated central portion. This method, together with circuitry which rejects pulses where charge is shared between the central and outer detector regions, reduces the background due to incomplete charge collection resulting in substantial improvement in the detectability of peaks in complex spectra. This technique is particularly important in the case of photon-excited X-ray fluorescence, where the scattered photons occur in the high-energy portion of the spectrum. A continuous distribution of degraded pulses resulting from these events would seriously limit the detectability for this type of measurement.

III. SIGNAL PROCESSING ELECTRONICS

A. Input Circuit-Preamplifier

1. Choice of Field-Effect Transistor

The signal processing electronics that amplifies and processes the weak detector signals is critically important in an X-ray spectrometer. The electronics is generally comprised of a preamplifier, placed in very close proximity to the detector, and a main amplifier which may be located some distance from the detector and its cryostat. The preamplifier usually contains a front-end amplifying element and associated components which are mounted near the detector inside the cryostat, while the remaining preamplifier components are placed as close as possible outside the vacuum wall of the cryostat. Modern high-resolution X-ray spectrometers employ selected field-effect transistors (FETs) which exhibit very low noise when operated at an optimum temperature in the region of 140°K. Since the detector is operated near 77°K to minimize its leakage current, the FET temperature must be raised slightly by providing a suitable heating to thermal resistance arrangement.

Selection of suitable low-noise FETs is a time consuming process, since very few samples of standard commercial FETs exhibit the required behavior at low temperatures. The inferior performance of most FETs is generally attributed to the presence of uncontrolled impurities which provide traps in the active region of the device where generation and recombination of holes and electrons can occur. Fluctuations in this process, as observed in the frequency range set by the amplifier bandwidth, are temperature-dependent and they contribute to a degradation in the energy resolution of the system.

Choice of a suitable FET is also based on matching the device to the detector capacitance. Input signals consist of charge produced in the sensitive region of the detector which flows in the external circuit as it is collected. Figure 7 shows the basic input circuit factors which contribute to determining the energy resolution of the spectrometer. The following parameters are important:

- Detector signal — Q
- Detector capacitance — C_D
- FET input circuit capacitance — C_{in}
- Detector leakage current — i_D
- FET gate leakage current — i_G
- FET channel noise represented by a series voltage noise source e_n
- The value of e_n is given by:

$$\overline{e_n^2} = 4KR_FT\Delta f \tag{6}$$

where K is Boltman's constant, T is the temperature (°K), Δf is the bandwidth of the system, and R_F is the FET noise resistance given approximately by

$$R_F = 1/g \tag{7}$$

where g is the transconductance of the FET. In the design of an FET, the transconductance of g and the FET capacitance C_{in} are linked and for practical purposes the ratio g/C_{in} can be regarded as constant.

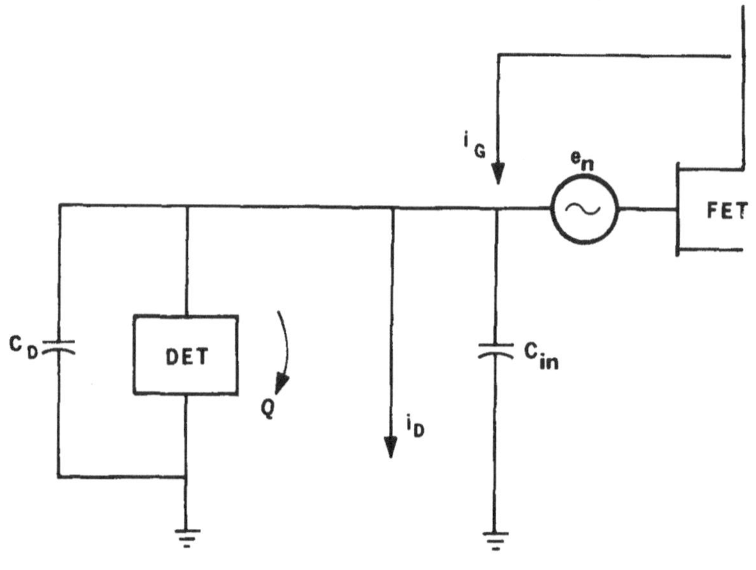

FIGURE 7. Noise sources in a semiconductor spectrometer.

As far as the FET channel noise is concerned, it can be represented by the generator e_n and, for signal to noise calculations, it is to be compared with the voltage signal $Q/(C_D + C_{in})$. The signal-to-noise ratio (for this noise source alone) is clearly proportional to $Qg^{1/2}/(C_D + C_{in})$ and therefore proportional to $QC_{in}^{1/2}/(C_D + C_{in})$. It is easy to show that this expression has its maximum value when $C_D = C_{in}$ and therefore that the FET should ideally be chosen to satisfy this condition. In this case, the signal-to-noise ratio will be proportional to $Q/C_D^{1/2}$. Clearly, the best resolution is obtained by using a small area detector, thereby minimizing C_D; however, the efficiency for detecting X-rays must also be considered, so, for typical X-ray fluorescence systems, a compromise of a 0.5 to 1 cm diameter detector is usually employed. This results in a detector capacitance ~1 pF. While the ideal FET, by these arguments, would be one having a similar capacitance, practical considerations have resulted in the use of FETs of the 2N4416 generic class having $C_{in} \simeq 3$ pF and g ~5 mA/V.

It is obvious from Figure 7 that e_n is not the only noise source, since fluctuations occur in both the detector and FET gate leakage currents. Furthermore, any additional parallel input components (such as a biasing resistor), would contribute noise. These parallel sources produce noise which, when expressed as an input voltage, is clearly proportional to $1/(C_{in} + C_D)$ as is the signal. Consequently, from the point of view of these noise sources, as distinct from the FET channel noise represented by e_n, the signal-to-noise ratio is independent of the total circuit input capacitance $(C_{in} + C_D)$.

2. Charge-Sensitive Method

The most obvious type of preamplifier would be a conventional voltage-sensitive amplifier. However, such a system would produce output voltage signals which would depend on the detector capacitance. This might result in rather serious drifts in signal amplitudes, since the surface channels on detectors change with time, contributing small changes in detector capacitance. Consequently, X-ray spectrometers all employ a feedback operational amplifier with a capacitive feedback element as shown in Figure 8, to produce an output signal whose amplitude is proportional to the charge flowing through the detector. Detector capacitance changes (within limits) then produce vir-

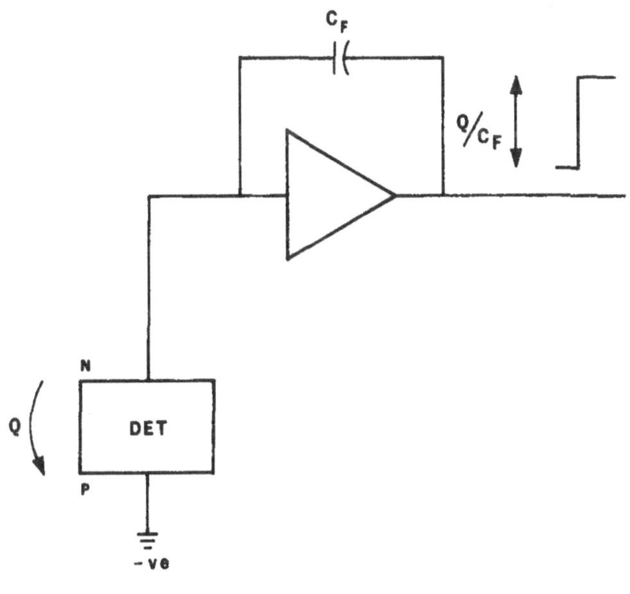

FIGURE 8. Charge sensitive preamplifier.

tually no change in the output signal (although they do affect the signal-to-noise ratio and, therefore, resolution). The output signal V_o from the charge-sensitive stage is given by:

$$V_0 = Q/C_F \qquad (8)$$

where C_F is the feedback capacity value.

3. Pulsed-Light Feedback

The simple operational amplifier integrator of Figure 8 would integrate the charge flowing in the detector resulting both from leakage current and radiation-produced ionization in the detector. Therefore, the output would drift in a positive direction with time and would soon saturate. Consequently, a method must be provided to discharge C_F and thereby maintain the preamplifier in its linear range. For many years, the method commonly used employed a high-valued resistor across C_F to provide a discharge path. This is far from ideal, since the resistor acts as a noise source and adds capacitance in the input circuit. Furthermore, high-valued resistors do not behave as pure resistors and their behavior causes difficulties in achieving good resolution at high-counting rates.

Most modern X-ray spectrometers employ a pulsed method to discharge C_F periodically. The pulsing may follow each event, or more commonly, may be initiated when the preamplifier output reaches an upper direct current level where a discriminator fires to start the charge restoration process. Charge restoration ends when the output falls to a predefined lower level. The pulse processing system is made insensitive during this reset time, so no inputs are processed during the few microseconds while C_F is being discharged. Since no direct current discharge current is present when signals are being processed, no noise is produced by the discharge process. This is in contrast to the resistor and dynamic charge restoration methods.

The most common pulsed restoration technique uses a pulsed light-emitting diode (LED) to cause substantial leakage current to flow in the FET drain to gate junction.[6] This is known as the pulsed-light or pulsed-opto feedback method. Since this is in very common use, we will describe it in more detail with the aid of Figure 9.

The detector is connected to the FET gate with negative detector bias applied to the

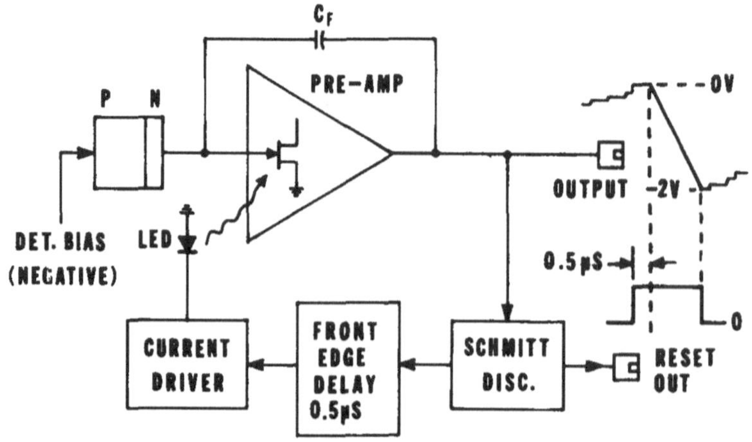

FIGURE 9. Details of a pulsed-light feedback preamplifier. The light-emitting diode (LED) is turned on when the output of the charge integrator exceeds the level set by the Schmitt discriminator.

p contact. Leakage current and photon-induced charge is integrated by the feedback capacitor C_F, causing the output to rise in voltage. When the output level crosses the upper trigger threshold of the Schmitt trigger circuit, the LED is turned on (after a slight delay). Light from the LED is directed onto the FET causing a substantial photocurrent to flow from drain to gate of the FET. This current, integrated by C_F, causes the output voltage to fall rapidly until it reaches the lower trigger threshold of the Schmitt trigger circuit. At this point, the light is immediately turned off. The Schmitt circuit is designed for a 2 V difference between its two trigger levels and the discharge of $C_F(\sim 0.1 - 0.2$ pF) through this voltage requires only a few microseconds. In normal use, the discharge cycle is only repeated every 10 to 100 msec so the loss of counts during the discharge is almost negligible. However, during the reset period, the preamplifier provides a blanking waveform which is used to inhibit baseline restoration at the output of the main amplifier during this period.

B. Amplifier Requirements

1. Gain, Linearity

Essentially, the preamplifier provides a very low impedance input to match the detector and a low output impedance to drive the cable to the main amplifier. It may also contain a certain amount of gain to increase signals and thereby to reduce the effect of any extraneous electrical noise picked up on this cable. Most of the gain in the system is provided by the main amplifier. The overall system gain requirement is easy to calculate. Since the detector is a source of signals consisting of a few hundred electrons (e.g., $\sim$300 for a 1 keV X-ray), and its capacitance is about 1 pF, typical output voltages for the charge integrator are in the range of 50 μV. An adequate output level from the main amplifier to the digital processing system is typically a few volts. Consequently, in this crude picture, the requirement is for a gain in the range of 10^5.

For many purposes, a rather precise linearity between input and output signal must be maintained, since nonlinearity of response would result in complications in interpreting spectra, particularly when computer processing is used. Generally speaking, the output must not deviate by more than 1% at any amplitude compared with its predicted value on the basis of a true linear response.

Another essential amplifier characteristic is the ability to adequately handle large overload pulses, (i.e., to recover from the overload quickly and without producing spurious output signals). This requirement is particularly important in pulsed reset systems, since a very large signal of the opposite polarity from normal signals results from the discharge of the feedback capacitor in the preamplifier.

2. Pulse Shaping and Its Effect on Resolution

The main amplifier also performs a very critical operation which is of great importance in determining the energy resolution of X-ray spectrometers. This operation consists of shaping the signal pulse to permit the best possible measurement of each signal amplitude despite the effects of noise of various kinds and of the possible interference of other (earlier) signal pulse residues. The topic of optimum signal processing is much too extensive for this short treatment of X-ray systems, so we will restrict this discussion to some rather basic features of the subject. The simplest picture of the pulse-shaping process is to regard it as a method of restricting the system bandwidth to a minimum value consistent with retaining the signal and thereby cutting out noise in the frequency regions not required to process the signal. This simple picture implies a low-frequency limit — therefore a differentiator — and a high-frequency limit — therefore an integrator. The existence of an integrator and differentiator implies that the shaping system converts the input signal, which is a voltage step (see Figure 8) into a pulse of limited rise time and duration. A short duration pulse is also essential to prevent interference between successive signal pulses, but the optimization of resolution generally requires longer pulse widths than is desirable from the point of view of avoiding interpulse interference. Therefore, some compromise must be made between resolution and performance at high-counting rates.

To simplify the understanding of pulse-shaping techniques, we will define a parameter that we term measurement time (T_m). In principle, this can be any characteristic time of the amplifier output pulse shape; for convenience, we will use the pulse width as measured at the points on the rising and falling edges corresponding to 5% of the full pulse amplitude. Figure 10 shows some characteristic pulse shapes and the relationships that exist between our defined measurement time (T_m) and other time parameters. Most pulse shaping circuits are passive — that is, they contain circuit elements which are fixed and do not change with time. The RC integrator, differentiator of Figure 10(A), is the simplest example. The long tail on this pulse is detrimental to the performance of a system at high-counting rates, so this type of shaping is rarely employed. The use of multiple RC integrators, as in Figure 10(B), results in more symmetrical pulse shapes. As the number of integrators is increased, a Gaussian pulse shape is approached. The symmetry of these pulse shapes leads to a good compromise between energy resolution and high-rate performance. Consequently, such pulse shapes (with >4 integrators), or slightly modified versions, are very commonly employed in X-ray spectrometers. Somewhat better performance can be achieved by some so-called active pulse shapes in which the values of components are switched in time, synchronous with the signal. A simple example is the gated-integrator shown in Figure 10(C), but more complex systems are slowly coming into use.

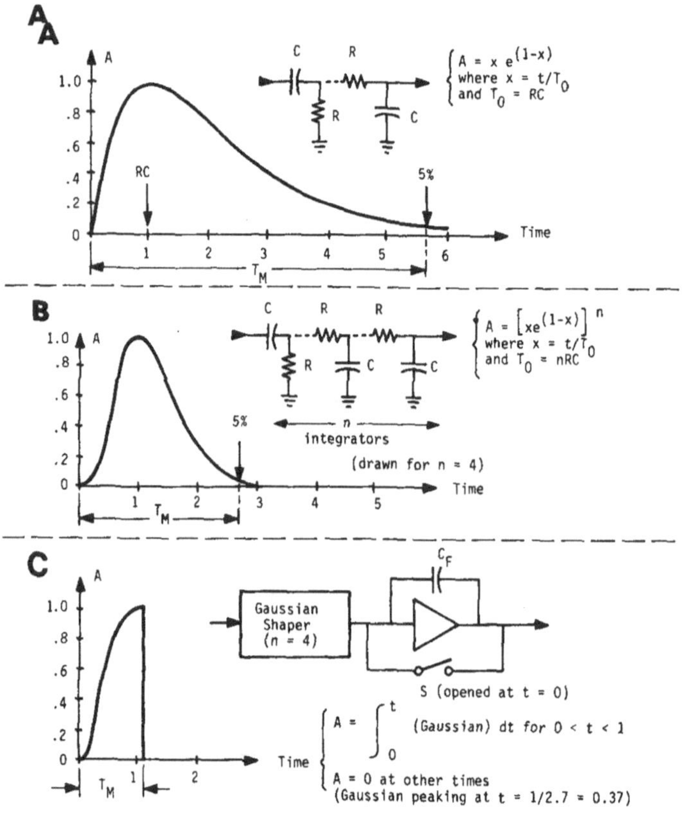

FIGURE 10. Some common pulse shapes used in X-ray spectrometer amplifiers. (A) RC differentiator — RC integrator. (B) RC differentiator — multiple RC integrator. (C) Gated integrator.

While the various types of pulse shapes achieve somewhat different energy resolutions and rate performance, once a type of shaper is selected, the characteristic times of the output pulse, including T_m, must be chosen. Since T_m largely controls both energy resolution at low-counting rates, and its degradation at high rates, choice of the optimum value of T_m for a given application is critically important. Many systems permit the user to adjust the measurement time to suit the application. The effect of measurement time on the electronic component of energy resolution is illustrated in Figure 11. The parameter Υ_o is the peaking time for a Gaussian shaped pulsed, $T_m \approx 2 \Upsilon_o$. As T_m is increased from a small time, we see that the resolution improves initially ($Res^2 \propto 1/T_m$), passed through an optimum value and then starts to rise ($Res^2 \propto T_m$). The short time part of the curve is dominated by channel noise in the FET (e_n in Figure 7), while the long time behavior results from the leakage noise (i.e., fluctuations in i_g and i_D of Figure 7). In the context of this discussion, energy resolution refers solely to the electronic noise contribution. Detector charge production statistics, as discussed in Section II(C), are not included. The depth of the valley in the curve in the region of the optimum value of T_m is determined partly by the absolute magnitude of these noise terms and also by miscellaneous noise sources (generally called 1/f noise — including insulators, surface noise in the FET detector, etc.), which produces a flat noise distribution independent of T_m.

Improvements in energy resolution over the past 10 years have largely been the result of achieving reduced 1/f noise, decreasing the leakage currents and their associated noise and of the use of pulsed light feedback which removes all noise due to currents discharging C_f in Figure 8. These improvements have all resulted in an increase in the optimum measurement time T_m, from the point of view of resolution. The use of measurement times as long as 100 μsec has made the problem of maintaining performance at high-counting rates more difficult. Improvement of the short measurement time part of the curves of Figure 11 depends on the possible availability of FETs with much better g/C_{in} ratios than present-day devices. While some progress is being made in this direction, any immediate change in the situation cannot be anticipated.

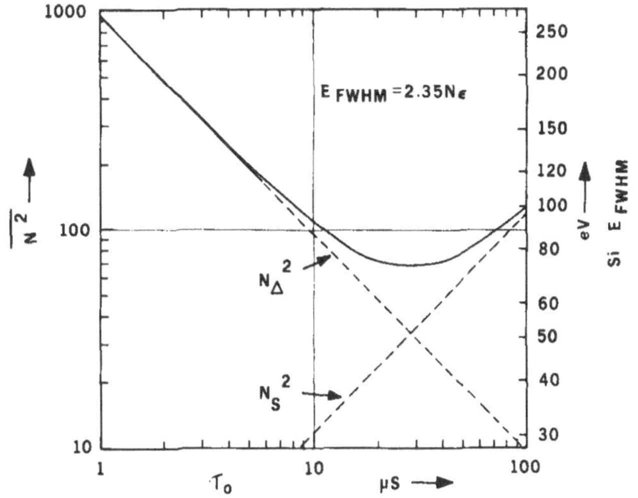

FIGURE 11. The effect of measurement time on resolution. Υ_o is the peaking time for a symmetrical Gaussian shape $T_m \approx 2\Upsilon_o$. The ordinate is given either in the number of hole-electron pairs (N^2) or in equivalent eV.

C. System Considerations

The previous discussion can best be focused by studying a complete X-ray spectrometer electronic system as shown in Figure 12. While other systems differ in many details, the general features must be provided in any modern spectrometer.

Signals from the preamplifier are fed to the input of this unit. The main linear amplifier block performs the functions of differentiation, amplification, and multiple integration to produce approximately Gaussian-shaped pulses at its output. This output is then fed to a baseline restorer which accurately defines the baseline levels of the signals; from this point through to the analog digital converter, which follows this unit, direct current connections are used and the baseline established by the restorer applies to all pulse amplitude measurements. Fluctuations in the baseline at high-counting rates will seriously degrade the energy resolution, so the restorer is designed to reduce such fluctuations to a minimum consistent with not significantly altering the basic signal pulse shape. Since the restorer normally clamps on negative voltage excursions, the very large negative signals which occur whenever the preamplifier resets would produce very serious effects on the baseline. Consequently, the clamping action of the restorer is inhibited during the preamplifier reset pulses and for a short time thereafter.

In addition to the main signal path, the preamplifier signal is differentiated to produce short signals which are amplified by the fast amplifier. A fast discriminator included in this block provides signals corresponding to the start of the Gaussian-shaped signals produced by the main amplifier channel. These fast signals are available for use externally (e.g., for timing purposes), but are also used internally for pile-up rejection. The pile-up rejection circuit inspects to find whenever the interval between two fast pulses is sufficiently short that two pulses will interfere with each other in the slow pulse-height measuring channel. If so, one or both (depending on the actual case) of the pulses are removed by preventing the signal discriminator from firing, thereby stopping transmission of signals through the stretcher to the output.

Signals from the baseline restorer output feed a stretcher circuit and if they exceed a lower level set by the discriminator and satisfy pile-up criteria (i.e., they are not contaminated by other pulses), they are stretched at their peak amplitude. The differential linear gate is then used to generate a rectangular signal output pulse equal in

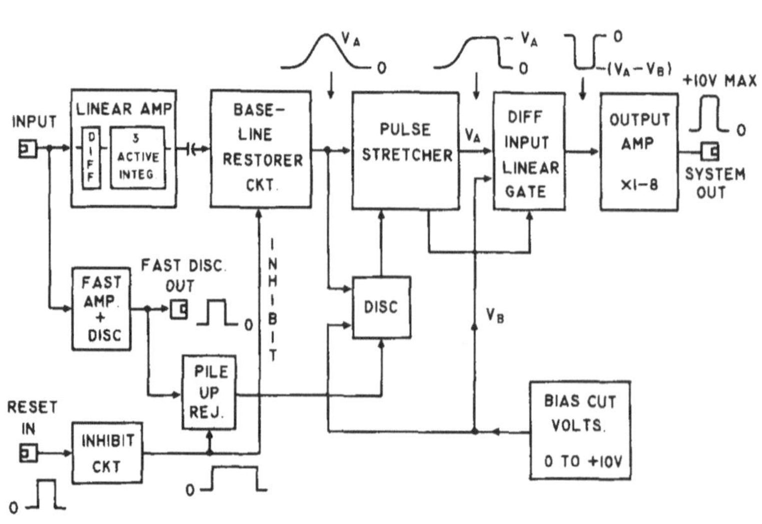

FIGURE 12. A complete X-ray spectrometer electronic system including baseline restorer and pile-up rejector.

amplitude to the signal peak height minus the discriminator bias level. Only that part of signals exceeding the bias level is passed on to the output amplifier, where signals are amplified to conform to the signal amplitude range of the pulse-height analyzer.

Dead time corrections must be applied in all quantitative applications of X-ray spectrometers. This is done by automatically adjusting the counting time to extend it by a factor sufficient to correct for those signal pulses lost due to system dead time. This is usually accomplished by counting a stream of accurately timed clock pulses gated by a dead time waveform generated by the pulse processing system. The resulting output clock pulses are a measure of the live time of the spectrometer. The dead time waveform must accurately represent losses due to the pile-up rejector process as well as any dead time in the pulse-height analyzer.

D. Pulsed-Excitation Technique

When using conventional continuous excitation sources and heavy samples, a condition where very high event rates occur in a detector, the pile-up rejector circuit of Figure 11 may reject almost all the pulses, so very few reach the analyzer. The relationship between the system output rate and input rate for a certain value of the pile-up inspection time is shown in Figure 13. It is seen that the output rate follows the input rate at low input rates, then falls for very high input rates. The peak output rate occurs at an input rate equal to 1/(Pile-up Inspect Time) and is 1/e times the input rate at this point. As discussed above, the optimum resolution in X-ray systems occurs at long measurement times and pile-up inspection times of 50 μsec or more are quite common. If we assume 50 μsec, the output rate is limited to about 7 kc/sec (i.e., 20/e).

A better output rate can be achieved if a pulsed excitation source, (such as a pulsed X-ray tube or a pulsed ion beam) is used in place of the dc excitation.[7] In a pulsed system, the beam is turned on at high level and turned off as quickly as possible when a signal is detected in the system (for example, by the fast discriminator of Figure 2.12). The excitation source is turned on again when the signal has been processed. In such a scheme, each pulse can be processed with almost no chance of pile-up since the excitation source is turned off during the pulse processing time. A small chance does

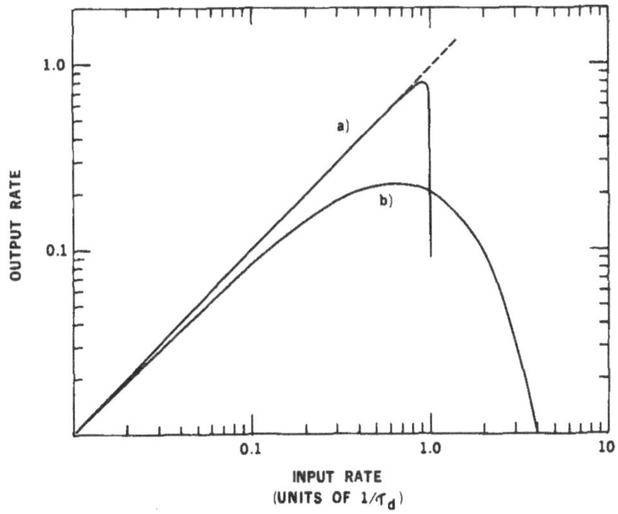

FIGURE 13. Output-counting rate verses input-counting rate. T_d is the system dead time and is normally slightly greater than the measurement time discussed in the text; conventional DC source, pulsed-excitation system.

exist of a second pulse arriving in the small time interval between the sensing of a signal and the excitation disappearing (an interval typically of only ~100 nsec). The input-to-output rate behavior of a pulsed system is shown in the second curve of Figure 13. It is seen that the input curves track very well until the rate reaches 1/(pulse processing time). Obviously, this rate cannot be exceeded under any circumstances, however this output rate is almost three times greater than can be achieved in a spectrometer using DC excitation.

The advantages of the pulsed excitation method are not achieved without some cost in complexity. It is necessary that the excitation source intensity should, as far as possible, be automatically adjusted for the system to operate near the peak of the curve of Figure 13. For heavy samples, this means reducing the source intensity while light samples require maximum power operation of the source. Therefore, the excitation intensity is not constant and it cannot be used as a reference for comparison of different samples that are counted for a fixed live time. Instead, the integral of the intensity must be measured and counting carried out for a fixed value of this integral. If a pulsed X-ray tube is used as the excitation source, the total charge through the tube is a good measure of the integral of the intensity. The complexity of these techniques prohibits their use except in situations where very large-scale analyses of samples are required and the maximum throughput is essential. The method has been applied and is almost essential to handle the volume of samples generated in large-scale studies.

IV. DIGITAL PROCESSING

The output of the electronic system shown in Figure 10 consists of a sequence of pulses of varying amplitude which reflect the energy distribution of photons in the fluorescence spectrum. In order to display and interpret this information, the amplitude of each pulse is measured, converted to a digital number, and stored in histogram fashion in a suitable memory array. The histogram pulse height spectrum is then available for subsequent examination and interpretation.

There are a number of commercial multichannel analyzer manufacturers who make instruments suitable for this operation. For typical X-ray fluorescence applications, a total of 1024 channels is used to encode pulses over an energy range of 30 keV or less. This implies that a 200 eV-wide Gaussian peak would be represented by seven channels above the half-maximum point of the distribution. This is more than adequate for most applications.

Commercial pulse height analyzers often include computerized facilities for locating X-ray peaks, calculating intensities, and applying calibration data to the fluorescent intensities. There are a number of methods employed for spectral analysis and for correcting the intensities for interferences from other X-ray lines and for X-ray absorption effects. The choice of method depends to a large extent on the nature of the sample studied and the degree of automation desired. More complete discussions of spectral analysis are available in the literature.[8]

REFERENCES

1. **Jenkins, R.**, *An Introduction to X-ray Spectrometry*, Heyden and Son, New York, 1974.
2. **Birks, L. S.**, *X-ray Spectrochemical Analysis*, 2nd ed., Interscience, New York, 1969.
3. **Leibhafsky, H. A., Pfeiffer, H. G., Winslow, E. H., and Zemany, P. D.**, *X-ray, Electrons, and Analytical Chemistry*, Wiley-Interscience, New York, 1972.
4. **Goulding, F. S. and Stone, Y.**, Semiconductor radiation detectors, *Science*, 170, 280, 1970.
5. **Goulding, F. S., Jaklevic, J. M., Jarrett, B. V., and Landis, D. A.**, Detector background and sensitivity of X-ray fluorescence spectrometers, in *Adv. X-Ray Anal.*, 15, 470, 1972.
6. **Landis, D. A., Goulding, F. S., Pehl, R. H., and Walton, J. T.**, Pulsed feedback techniques for semiconductor detector radiation spectrometers, *IEEE Trans. Nucl. Sci.*, NS-18(1), 115, 1971.
7. **Jaklevic, J. M. and Goulding, F. S.**, Semiconductor detector X-ray fluorescence spectrometry applied to environmental and biological analysis, *IEEE Trans. Nucl. Sci.*, NS-19(3), 384, 1972.
8. **Jaklevic, J. M., Loo, B. W., and Goulding, F. S.**, in *X-Ray Fluorescence Analysis Of Environmental Samples*, Ann Arbor Science, Ann Arbor, Michigan, 1977.

Chapter 3

EXCITATION METHODS FOR ENERGY DISPERSIVE ANALYSIS

Joseph M. Jaklevic

TABLE OF CONTENTS

I. Introduction 49

II. General Features of X-Ray Fluorescence Excitation 50
A. Photon Excitation 50
B. Electron Excitation 53
C. Heavy Charged Particles 53

III. X-Ray Tubes and Radioisotope Sources 57
A. X-Ray Tubes 57
B. Radioisotope Sources 59

IV. Charged Particles from Accelerators, Protons, and Alphas 60

V. Electron Beams: Microprobes and Scanning Electron Microscopes 63

VI. Special Topics 64
A. Proton Microprobe 64
B. Polarized X-Ray Sources 65
C. Synchrotron Radiation 65
D. Low Energy X-Ray Detection 65

References 66

I. INTRODUCTION

Many of the recent advances in the area of energy dispersive X-ray fluorescence analysis (XRF) have been the result of innovations in the manner in which fluorescence excitation is achieved. Traditional XRF methods employed X-ray tubes to produce continuum photon distributions which irradiated the sample to be analyzed. The resultant fluorescent spectrum was then analyzed using a very high resolution Bragg crystal spectrometer to sequentially measure the energy and intensity of characteristic X-ray lines.

The advent of semiconductor detector X-ray spectrometers presents new opportunities to improve upon the traditional wavelength dispersive X-ray fluorescence method. The high efficiency with which semiconductor detectors can measure a range of characteristic X-ray energies allows the use of less intense excitation sources. On the other hand, the somewhat poorer energy resolution of these detectors compared to crystal spectrometers requires that greater care be taken in maximizing the fluorescent signal relative to unwanted background X-rays. By careful consideration of these factors, it is possible to develop powerful analytical methods.

Some of the more important X-ray excitation methods which have become widely used in recent years include low-power X-ray tubes,[1,2] radioisotope[3,4] sources, polarized X-ray beams,[5-7] fine-focus electron-beam probes,[8,9] and heavy charged particle beams from accelerators.[10,11] Each method has particular advantages and disadvantages which determine its use in a particular application. The choice of a method depends on a number of factors including availability, cost, speed, accuracy, and detectability. In the present chapter, we will attempt to describe the various methods in sufficient detail so that a meaningful comparison can be made.

II. GENERAL FEATURES OF X-RAY FLUORESCENCE EXCITATION

The fluorescence process can be visualized with the help of Figure 1. The incident radiation from the excitation source interacts in the atoms of the sample and, in some cases, creates vacancies in the inner atomic shell of the element of interest. These vacancies in turn can de-excite with the emission of X-rays whose energies are characteristic of that particular atom. To the extent that these atoms are evenly distributed in a homogeneous matrix and provided that the incident radiation flux is uniform, the number of X-rays detected in the spectrometer is proportional to the concentration of that element.

In general, the probability that a particle in the incident radiation flux will interact with a given trace constituent and produce a useful fluorescence X-ray is rather small. Most of the excitation radiation continues through the sample until it interacts with the atoms which constitute the bulk of the sample matrix. These interactions often produce unwanted background X-rays which interfere with the measurement of fluorescence X-rays. In biological samples, the sample matrix is normally dominated by the lighter elements typical of hydrocarbon compounds. The interaction of the incident flux in the matrix can give rise to secondary X-ray distributions which constitute a continuous background in the fluorescence spectrum. The ideal excitation source produces a minimum amount of the unwanted background radiation while at the same time efficiently and uniformly fluorescing those atoms in the atomic number range of interest. Since no method has proven ideal for all types of applications, the choice of an optimum excitation method involves an understanding of how these various factors operate in the case of each type of incident radiation normally employed in XRF.

A. Photon Excitation

This technique involves the use of electromagnetic radiation (X-rays) with energies sufficient to create vacancies in the atoms whose concentration is to be measured. The radiation is normally obtained from either X-ray tubes or radioisotopes sources. In the former case, the radiation is the result of the interaction of an energetic electron beam with the anode material of the X-ray tube. In radioisotope sources, the radiation

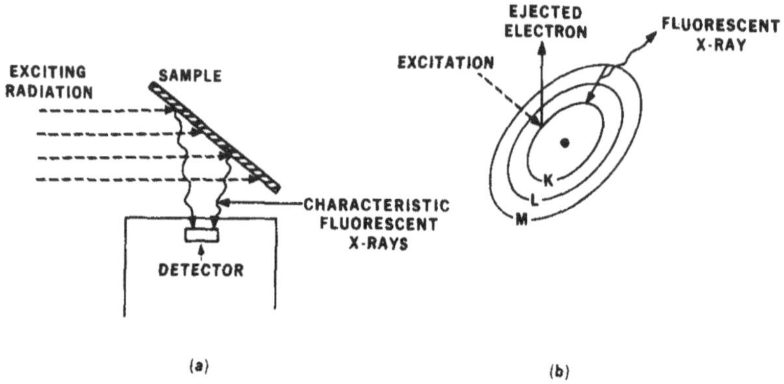

FIGURE 1. Schematic diagram of the energy dispersive X-ray fluorescence method. Figure 1a shows geometry of excitation, sample, and detector; Figure 1b illustrates a simplified picture of the fluorescence process.

is due to gamma-ray or X-ray transitions which result from the radioactive decay process. The incident X-ray spectrum can be either continuously distributed over a broad energy spectrum or concentrated in a discrete series of lines.

Photons incident on the sample interact via the photoelectric effect to produce inner-shell vacancies which give rise to the fluorescent lines. The probability for vacancy production is greatest for photon energies immediately above the threshold for the ejection of the bound electron. The probability then decreases as the inverse cubic power of the energy of the incident photon as it is increased above this threshold. This energy dependence has important consequences when one wishes to analyze a wide range of elements using a single energy photon source since the sensitivity will vary proportionally.

The probability for vacancy production is normally expressed in terms of a cross-section per atom which can be interpreted as the effective area of that particular atom when measured by the interaction with a beam of particles. Figure 2 is a plot of the photoelectric cross section for a number of elements of interest as a function of photon energy.[12] It is important to note that, for a given incident energy, the cross-section for vacancy formation varies significantly for different elements. For example, using 10 keV photons, the probability for excitation of an Fe atom is 6.6 times greater than for Al. This represents a factor of 20 on a mass basis since Fe is more dense.

In order to overcome this wide variation, it is sometimes desirable to either use a continuum of X-ray energies in the excitation spectrum or perform a series of fluorescence measurements using several discrete energy photon sources which span the atomic number range of interest.

The sensitivity of analysis with photon excitation depends not only upon the probability of fluorescence excitation, but also on the relative probability for producing unwanted radiation which can interfere with the observation of the lines of interest.

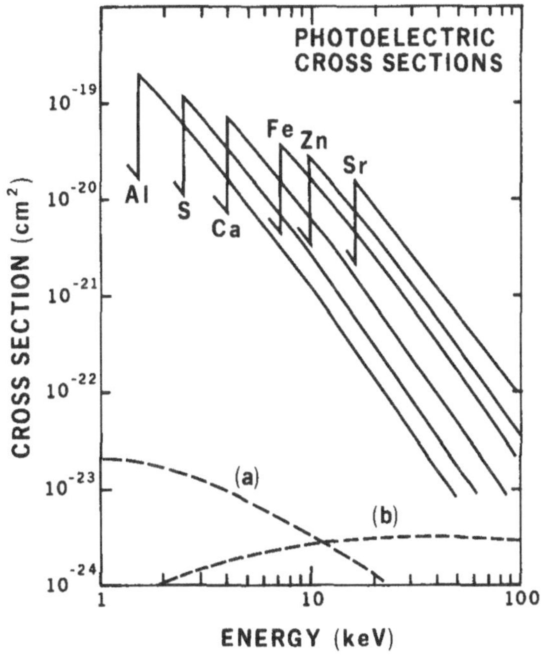

FIGURE 2. Cross section of K-shell vacancy production using photons for a few elements. The dashed curves show scatter cross sections for carbon. (a) is elastic; (b) is inelastic.

For samples consisting predominantly of lighter elements, the most serious problems arise from the scattering of the incident photons in the sample matrix. In addition to the photoelectric effect which is exploited to produce the characteristic fluorescence, photons can also interact in matter via elastic and inelastic scattering processes. In elastic scattering, the photons' direction of propagation is changed without any loss in energy. Inelastic scattering results in a change of direction and a loss in energy which is governed by the well-known Compton equation. In either case, the photons in the incident beam can be scattered into the detector resulting in unwanted events in the pulse-height spectrum.

The probability for such scattering events can likewise be described by a cross-section such as the dashed curves shown in Figure 2 which are measured for the case of carbon. Curve (a) is the elastic scattering cross-section, (b) is inelastic. Although these cross-sections are several orders of magnitude less than the photoelectric cross section, the hydrocarbon matrix is normally present in abundance far greater than the trace element of interest. Choosing 10 keV radiation on a sample of 10 ppm Fe by weight in a carbon matrix, the ratio of scattered to fluorescence X-rays would be approximately 500 to 1.

A better idea of the effect of scattering on the fluorescence measurement can be obtained by examining the idealized spectrum shown in Figure 3. Here we consider a monoenergetic photon source incident upon a biological sample consisting of a light-element matrix containing several trace elements. The prominent peaks are due to scattering in the matrix; the highest energy peak represents elastic scattering; the other, broader peak reflects the angular spread over which the inelastically scattered photons were detected. The smaller peaks in the middle of the spectrum represent the fluorescent X-rays of interest.

The detectability of the spectrometer is determined by the ratio of the fluorescence peaks to the background beneath them. In this illustration, the background below this peak is due primarily to incomplete collection of the energy deposited by the large scatter peaks for some small fraction of events. Since this background is proportional to the scattered intensity, it is obvious that the sensitivity of the measurement is limited by matrix scattering of the incident radiation.

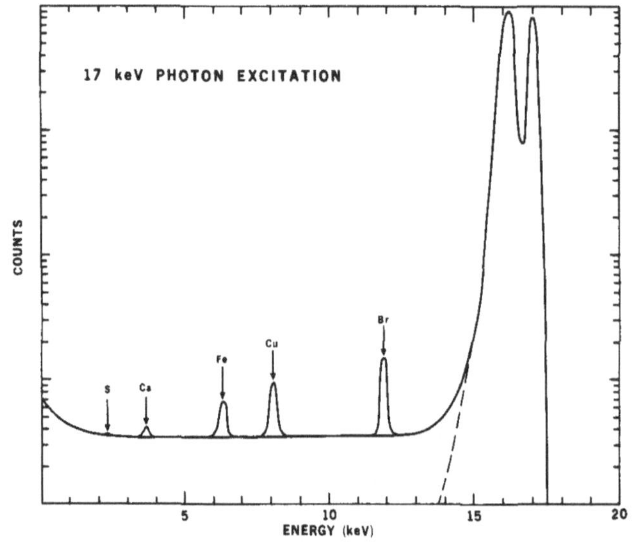

FIGURE 3. Schematic pulse height spectrum of a sample fluoresced with monoenergetic photons. The sample corresponds to elements present at the 10 ppm level in a hydrocarbon matrix.

As noted earlier, the variation in sensitivity due to the energy dependence of the photoelectric cross-section could be reduced by using a continuum of photon energies in the excitation spectrum. However, since a continuum excitation source would yield a superposition of spectra similar to that of Figure 3 integrated over all energies in the incident photon distribution, it is obvious that a much higher background would result. Figure 4 is a spectrum calculated for the same sample, but with continuum excitation such as might be obtained from an X-ray tube. Two extreme cases of scattering probability are shown. The more uniform energy dependence of fluorescence intensity is achieved only with an accompanying loss in the minimum detectable amount of trace elements.

Most systems which attempt to achieve lowest possible detectable limits use a series of discrete excitation energies to cover a wide range of elements.[13] The exception to this rule is the very low energy region where comparable sensitivities have been observed with continuum excitation. A comparison of results obtained with several types of photo-excitation methods is contained in Reference 14.

B. Electron Excitation

Electron beams are attractive candidates for fluorescence excitation because of the ease with which they can be generated at energies appropriate for vacancy production. Conventional high-voltage supplies can be used with simple thermionic electron sources to produce milliampere beams at energies up to 100 keV or more. Virtually all of the X-rays used in biological applications are the result of electron beam interactions with X-ray tube anodes.

Electrons incident on a target interact principally by an electrostatic interaction with the orbital electrons of the target atoms producing ionization. The probability of such interactions is large and results in the electrons loosing their entire energy in a very short distance. For example, a 50 keV electron has a range of 5×10^{-4} cm in a W anode. Although the number of ionization collisions within this distance is very large, most of them occur in the outer electronic shells where the fluorescence energies are well below the X-ray regions.

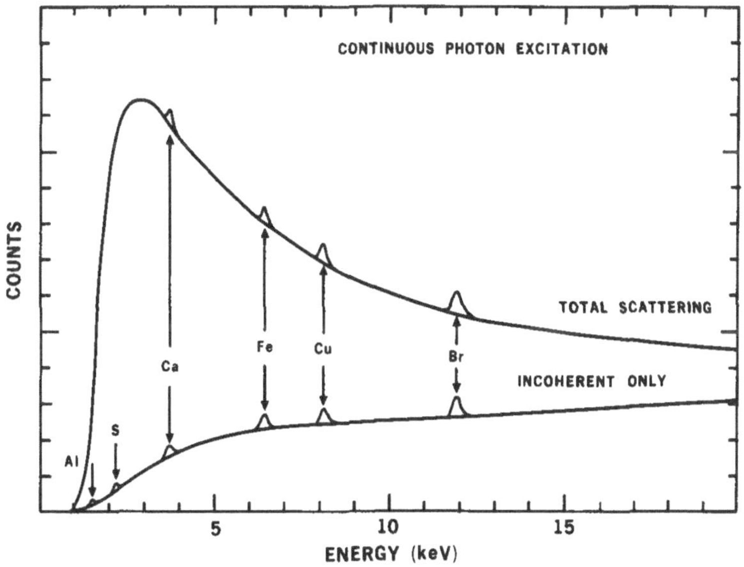

FIGURE 4. Schematic pulse height spectrum of the same sample as used in Figure 3, but assuming continuum photon fluorescence.

Figure 5 is a plot of the cross-section for K-shell ionization as a function of electron energy for several elements.[15] The cross-sections exhibit a less abrupt threshold than that for photons. The maximum cross-section is achieved at an incident electron energy 3 to 4 times the K-shell binding energy. It is interesting to note that the magnitude of these cross-sections are of the same general order of magnitudes as those shown for the photoelectric effect in Figure 2.

In addition to the ionization process, electrons interact in matter via a continuous deceleration process which produces a continuous X-ray background. The deceleration occurs when an electron is deflected by the positive field of a nucleus and results in a continuous distribution of emitted X-ray energies up to the energy of the incident electron. This "brehmsstrahlung" spectrum is a serious limitation on the use of electrons for trace analytical work since it is produced in interactions with any element in the sample including the matrix. The spectrum can be described approximately by the expression:

$$dN(E) = 2.76 \times 10^{-6}\ Z\ \frac{(E_0 - E)}{E}\ dE \qquad (1)$$

where dN is the number of photons between energies E and E + dE, and Z is the atomic number. A detailed examination of the signal/background ratio for electron excitation can be carried out using Equation 1 together with appropriate electron range and ionization cross-section data.[17]

Figure 6 is a schematic spectrum for the same hypothetical sample as was used in Figures 3 and 4, but using 20 and 40 keV electrons. A quantitative comparison of this spectrum with the equivalent one for photon or heavy charged particle excitation demonstrates the poorer peak-to-background ratio evident in the electron excited spectrum. Electron beam excitation finds its most important X-ray fluorescence application when used in microprobe or scanning electron microscope applications. Here the relatively

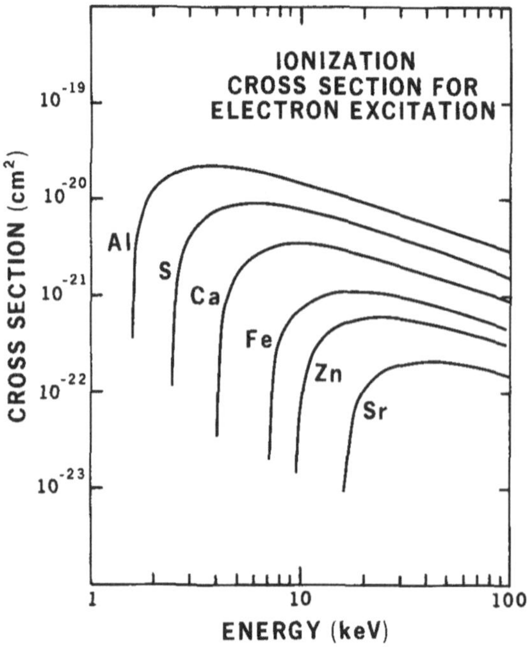

FIGURE 5. Cross section for K-shell vacancy production using electrons.

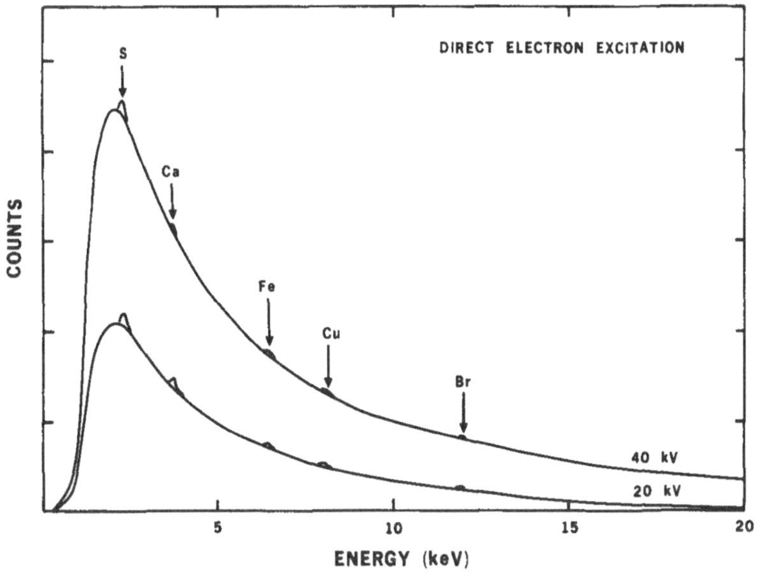

FIGURE 6. Schematic spectrum of the same sample used for Figures 3 and 4, but using direct electron excitation.

poorer sensitivity per unit volume is more than compensated by the ability to analyze very small areas. Since elements are not uniformly distributed in many types of sample, the ability to excite extremely small volumes is a powerful analytical tool.

C. Heavy Charged Particles

Recently there has been an increase in the use of heavy charged particle beams such as protons (H^+) and alpha particles (He^{++}). Heavy charged particles exhibit large ionization cross-sections together with a much lower background relative to electrons. These beams are more difficult to produce than other forms of excitation; nevertheless, a number of facilities capable of generating several MeV beams of such particles are currently available for use in XRF and are supporting vigorous programs. Many of these programs were discussed at a conference.[10]

The ionization cross-sections for heavy particle excitation are similar to that of electrons with equivalent velocities. The property which makes their use attractive in XRF is the virtual absence of direct brehmsstrahlung rays. The mass of the nuclear particles relative to electrons reduce the effect of acceleration due to interactions with sample nuclei, and hence the probability of brehmsstrahlung production. The cross-section for continuum ray production is in fact proportional to the reciprocal of the square of the projectile masses, i.e., a proton produces 10^{-6} times less brehmsstrahlung than an electron. This results in a greatly enhanced peak-to-background ratio relative to electron-beam excited spectra.

Without the continuum background to obscure fluorescence peaks, it would appear that detectable limits for heavy charged particle excitation would be vastly reduced relative to other methods. However, a second order process involving continuum radiation production by the energetic electrons ejected in the ionization process increases the detectable limit to a value similar to that of photon excitation.

Figure 7 is a plot of ionization cross-section as a function of energy for protons. The cross-sections are again comparable to those observed for other excitation methods, but with a somewhat different energy dependence. The background sources can be understood by noting that the ionization cross-section is larger for light elements. For biological samples, the majority of ionization events occur in a hydrocarbon ma-

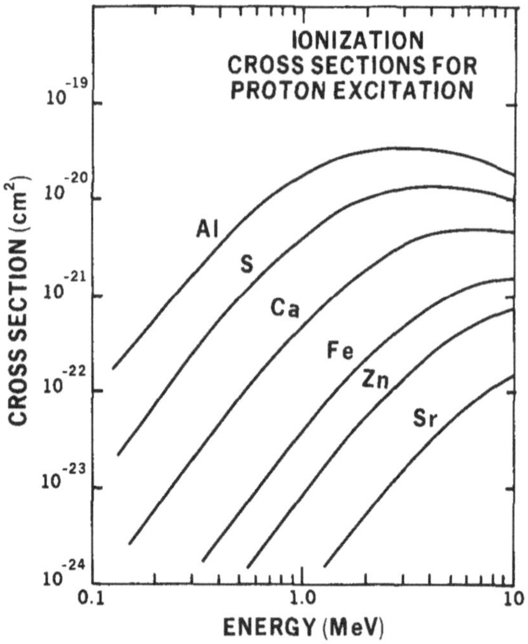

FIGURE 7. Cross section of K-shell vacancy production using protons.

trix where the cross-sections are much greater than in the trace constituents. The maximum energy of the ejected electrons is approximately equal to the beam energy times the ratio of the mass of the electron to the heavy particle mass. These ejected electrons produce continuum X-ray radiation as they slow down in the matrix material as discussed earlier. This background constitutes the major limit of detectability for charged particle analysis. However, since it involves a second-order process, the magnitude of the problem is far less severe than for direct electron excitation.

The spectrum in Figure 8 is a calculated distribution using the same sample form as discussed in the previous comparisons. The continuum distribution is calculated assuming an ejected electron distribution appropriate for 3 MeV protons. The solid and dashed curves assume slightly different models for the continuum X-ray production and detection. What is most important are qualitative differences between this and the proceeding spectra. The shape of the background is similar to the continuum distribution for electron excitation, but with a lower end point energy and reduced intensity. The peak-to-background ratio is comparable to photon excitation, but exhibits a different dependence on X-ray energy.

Inherent in the use of charged particle beams is the necessity for an electrostatic accelerator or cyclotron. Although it would not seem economical to build such a machine for analytical purposes only, there are a number of installations where these facilities exist and are available for part-time use in analytical programs.[10]

In addition to the inherently good sensitivity achieved with charged particles, there are numerous ways in which the focusing properties of the particles themselves can be used to advantage. For large area samples, it is possible to diffuse the beam over the entire sample resulting in a uniform excitation intensity. For small samples, the beam can be focused to illuminate a small area or, in extreme cases, beams as small as 1 μm can be generated for use as microprobes.[19] One disadvantage is the need to operate with the sample in vacuum for most applications. However, there have been successful attempts to generate external proton beams for some applications.[20] By passing the

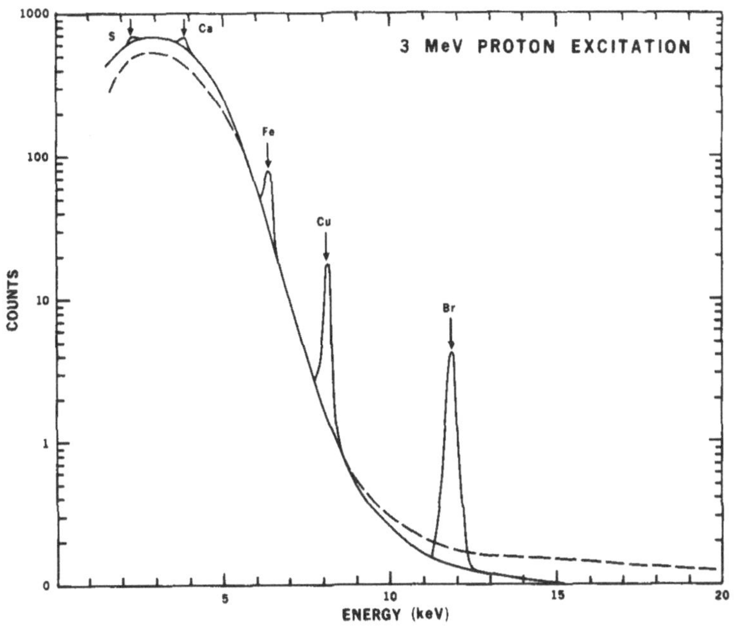

FIGURE 8. Schematic spectrum of hypothetical sample used previously, but using 3 MeV proton excitation.

beam through a thin foil into an He atmosphere at ambient pressure, it is possible to reduce target heating and charging effects and eliminate the inconvenience of vacuum operation.

III. X-RAY TUBES AND RADIOISOTOPE SOURCES

A. X-ray Tubes

The most intense photon sources which are generally available are X-ray tubes. A beam of 50 kV electrons impinging on a target will generate approximately 10^{-5} photons per incident electron — for a 10 mA electron current this represents 6×10^{11} photons/sec or 20 Ci. These photons are distributed between the characteristic X-rays of the anode material and the continuous brehmstrahlung distribution. This photon distribution can be used either directly or modified by external absorbers in order to produce more nearly monoenergetic radiation. The simplest way to produce a discrete energy spectrum is to place an X-ray transmission filter consisting of a thin (2.5×10^{-3} cm) foil of the anode material between the anode and the target. Since the characteristic X-rays of a given element have an energy immediately below the respective absorption edge energy, a foil of that element is a good transmission filter for those X-rays. A typical situation would be a Mo anode tube with Mo filtering resulting in a Mo $K_{\alpha,\beta}$ spectrum. This is a favorable situation for K-shell excitation of elements from Ca to Sr in the periodic table and for L-excitation of heavy elements such as Pb and Hg.

A more versatile scheme for obtaining discrete energy photon sources makes use of secondary fluorescence targets to transfer the output flux into the characteristic X-rays of the secondary target. Those X-rays in the tube output which are above the absorption edge of the secondary target are photoelectrically absorbed. The vacancies they create then deexcite with the emission of characteristic X-rays. Since the emitted X-rays are spatially isotropic there is always a sizeable loss of photon flux in this two-step process. However, the use of external secondary targets allows one to easily change the excitation energy in order to optimize the sensitivity for measuring the element of interest.

The advantages of multiple excitation energies in photon excitation can be seen in Figure 9. This is a plot of the relative efficiency for fluorescence X-ray production for a series of three fluorescent X-rays. The advantage of having the incident energy immediately above the absorption edge of the element of interest is apparent. It is also possible to select against one element by choosing an excitation energy which is below the threshold for excitation.

Due to the inefficiency of the secondary fluorescence process, it is performed either in a very close geometry where the respective detector-sample-secondary target distances are small or, alternatively, a more powerful X-ray tube must be used. Reasonable counting rates have been achieved with less than 100 W X-ray tube anode dissipation in the close-coupled case. X-ray tube powers of 1 kW or more can easily be achieved in the latter case if necessary.

Because of the normally isotropic distribution involved in X-ray production processes, X-ray excitation is well suited for large area samples, i.e., 1 to 10 cm^2. The ideal sample for X-ray excited fluorescence analysis is a thin, uniform, and homogeneous sample. The thinness minimizes absorption effects particularly for lower energy fluorescent X-rays. Uniformity and homogeneity ensure accuracy in the analytical results and facilitate corrections for absorption effect when necessary.

To the extent that the sample form departs from this idealized form, it may become necessary to perform corrections for matrix-absorption and particle-size effects. These both relate to the absorption of fluorescence X-rays due to thick, inhomogeneous, or irregularly shaped samples that are problems normally encountered in X-ray analysis

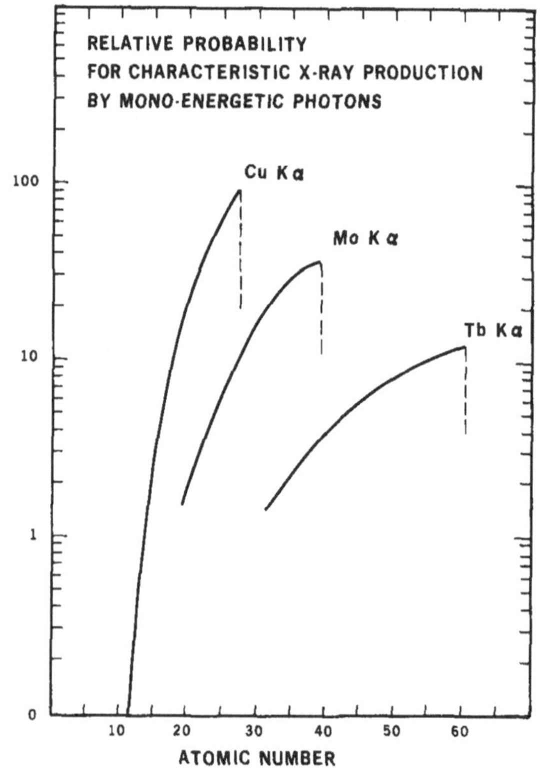

FIGURE 9. Relative probability for characteristic X-ray production for some typical secondary X-ray targets.

and extensive literature has been developed pertaining to methods for approximately correcting these effects.[21-23]

A number of intercomparison studies have established the validity of elemental analyses performed by this method.[24-26] In normal everyday useage, it is possible to attain precisions and accuracies of analyses of 5% or less. In favorable cases, even better results are possible. The long-term calibration stability of such instruments is excellent and can easily be within the above precision.

B. Radioisotope Sources

An alternative source of photon excitation is achieved through the use of radioactive isotopes. There are a number of isotopes which emit a limited number of X-ray or gamma-ray lines suitable for fluorescence excitation. Their advantage derives from the cost and convenience of use. However, since the maximum source intensity easily available is of the order of 1.0 Ci, the fluorescence intensity is less than that achieved with X-ray tubes. In situations where thicker samples are available for analysis and where longer analysis times can be employed, the use of radioisotope sources can be advantageous. Table 1 is a partial list of suitable radioisotope sources and the associated radiation energies. Many of these sources are commercially available in geometries appropriate for XRF analysis.

The same criteria for sample form which applied in the case of X-ray tube sources apply also for radioisotopes. Particle size and matrix effects are also encountered and may be more difficult to handle due to the diffuse geometry of the source and the close source-detector geometry. Figure 10 illustrates typical geometries employed in analysis. Figure 10A shows the case where the sample is directly excited by the radioisotope; 10B is a secondary-fluorescence geometry for variable energies. There have been attempts to exploit large area sources in order to produce a more uniform illumination of the sample and thereby reduce particle size effects. The inherent stability of isotopic sources has the potential advantage of better precision and accuracy although in practice this advantage is seldom exploited.

Figure 11 shows spectra obtained from an identical sample of NBS orchard leaves using radioisotope excitation and using Mo X-ray tube excitation.[27] The spectra are similar although the X-ray tube spectrum exhibits a higher background due to scattered brehmsstrahlung. The radioisotopes used in the measurements were $Cd\text{-}^{109}$ and $Fe\text{-}^{55}$. The difference in light element detectabilities using the two different excitation energies is qualitatively evident from the spectra.

It should be mentioned that radioisotope alpha-particle sources can be fabricated for certain applications. The intensity of such sources normally limits their use to the analysis of very light elements where the probability for ionization is large. An instrument using such a source has been constructed for use in the surface analysis of samples for very light elements. Figure 12 is a spectrum of a carbon film on a glass substrate obtained with an alpha-particle source operated in the same vacuum enclosure as the sample and detector.[28]

TABLE 3.1

Radioisotopes for X-ray fluorescence

Nuclide	Half-life	Emission energies
Fe^{55}	2.7 years	5.9 keV
Cd^{109}	453 years	22.1 keV, 87.7 keV
I^{125}	60 days	27 keV
Am^{241}	458 years	12-17 keV, 60 keV
Co^{57}	270 days	6.4 keV, 122 keV, 144 keV
Pu^{238}	86.4 years	12-17 keV

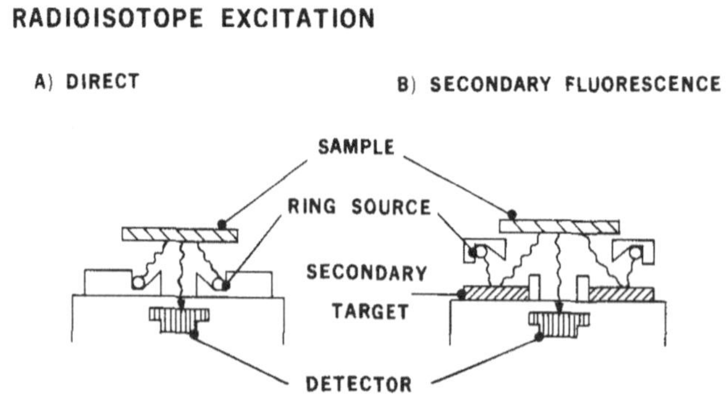

FIGURE 10. Geometries employed in radioisotope excitation.

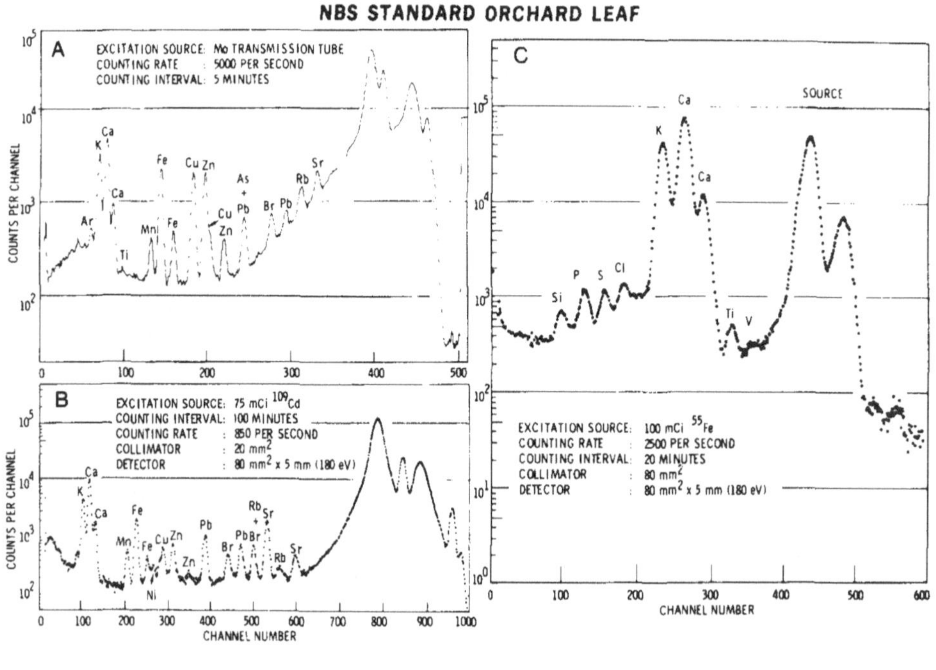

FIGURE 11. Spectra of NBS orchard leaf standard taken with (A) Mo X-ray tube excitation, (B) Cd^{109} radioisotope source, and (C) Fe^{55} source.

A number of references to energy-dispersive analysis using photon excitation are available.[29-31]

IV. CHARGED PARTICLES FROM ACCELERATORS, PROTONS, AND ALPHAS

Although the use of charged particle accelerators permits the use of a wide variety of projectile types and energies, developments in the field have centered on two principle methods of analysis. These typically use 3 MeV protons and 20 MeV alpha particles. An acronym commonly employed to describe these methods is PIXE or particle-induced X-ray emission analysis. The energies are chosen so that the endpoint of the ejected electron distribution results in a continuum distribution which does not interfere strongly with the X-ray region of interest.

Figure 13 is a diagram of typical charged-particle excitation apparatus.[22] The accelerated beam is directed down an evacuated beam line toward the target. Depending

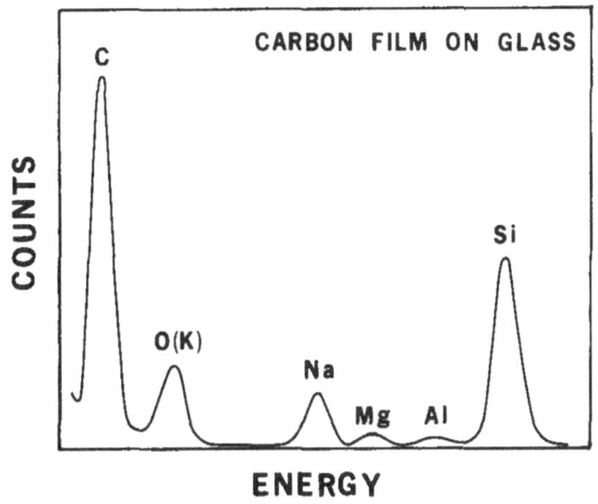

FIGURE 12. Spectrum of carbon film on a glass substrate obtained with an alpha-particle radioisotope source operated in the same enclosure as the sample and detector.

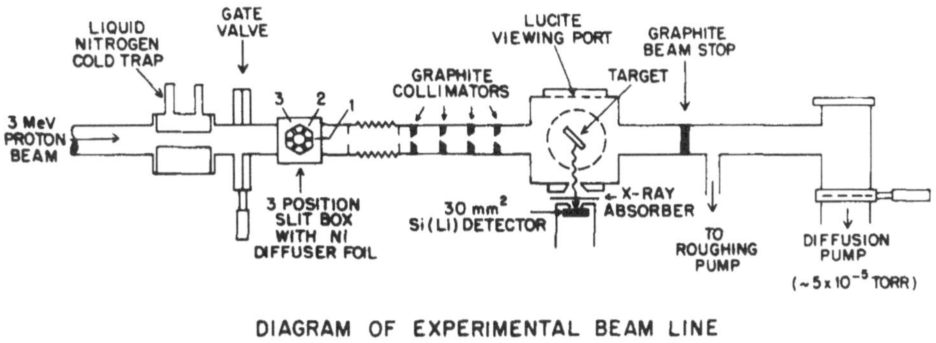

FIGURE 13. Diagram of a typical beam line for charged particle analysis. This particular system is in use at the Duke University.

upon whether a line focus or diffused beam is required, adjustments in the magnetic focusing element upstream can be performed. A diffuser foil is employed before the sample in order to uniformly distribute the particles over the sample.

The samples are normally contained in multiple target holders to reduce the frequency of entry into the vacuum chamber. The beam passes through the thin samples and is stopped in an absorber shielded from the detectors field of view. In the case of Figure 13, this is a graphite beam stop. The total current can be carefully measured and the result used to calibrate the yield of X-rays in the sample. X-rays are detected through a thin Be window in the side of the chamber.

In order to achieve optimal sensitivity in analysis, thin substrates are used whenever possible. The range of 3 MeV protons in carbon, for example, is approximately 16 mg/cm^2, and the ionization process is most effective over the early part of the energy loss process. The thin substrate limits the amount of beam current which can be directed on the sample since local heating in the nonconducting vacuum surroundings can evaporate the more volatile constituents or physically damage the sample. The use of conductive substrates such as thin carbon foils reduces the problem and also eliminates charging effects on the sample.

Figure 14 shows a series of spectra obtained on the NBS orchard leaf sample using a variety of excitation methods.[27] The reduced sensitivity for higher energy particles is

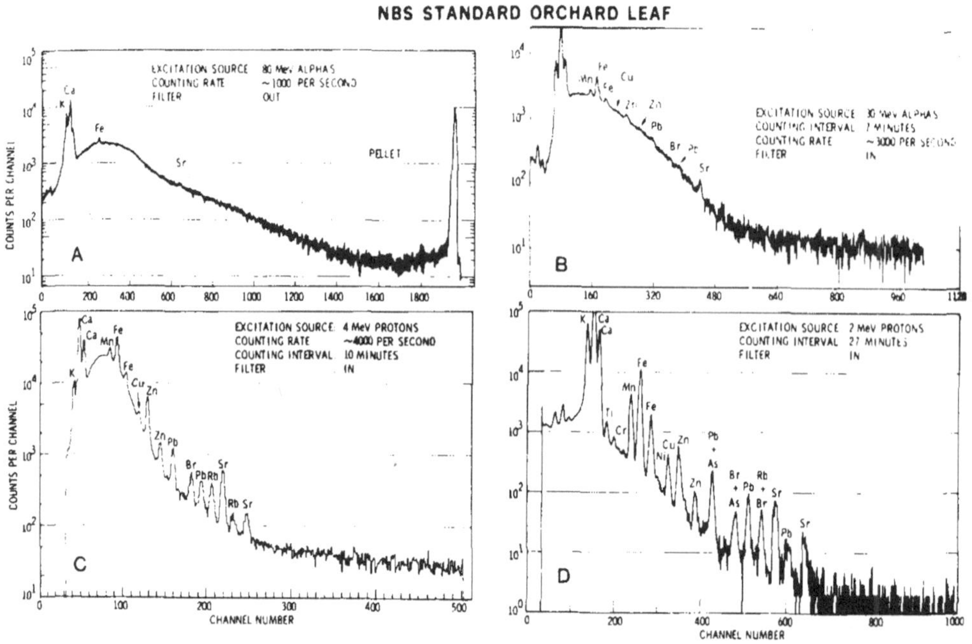

FIGURE 14. Comparison of spectra obtained on NBS orchard leaf standard using several choices of charged particle excitation. Most workers commonly employ either protons in the 2 to 3 MeV range or alpha particles of 15 to 20 MeV.

primarily the result of the higher endpoint energy for the continuum distributions. Most workers commonly employ protons at 3.5 MeV or less and alpha particles at 20 MeV or less.

Figure 15 is an example of several spectra of a bovine liver sample acquired at a series of incident proton energies. The differences between the two low energy spectra are due to the insertion of an absorber between the sample and detector to reduce the low energy counting rate and enhance the sensitivity for higher energy X-ray detection.

Types of samples which have been extensively studied by this method include thin-membrane, air-particulate filters, ashed biological samples deposited on thin backings, thin sections of lyophilized tissues, and other similar types of samples.

The best accuracies and sensitivities are achieved with thin samples ($<1/mg/cm^2$). For thicker samples, corrections due to the energy loss of the incident protons and absorption of the fluorescence X-rays must be included. For thick biological specimens, electrostatic charging of the sample and possible evaporation of volatile contituents in the vacuum must be considered. Willis et.al.[33] discuss these problems and present comparisons of analyses obtained on standard materials prepared as both thick and thin samples.

The precision and accuracy obtained with charged particle excitation can be made comparable to other XRF methods in most cases of interest. Cablibration is normally performed using thin films of known weight to establish the sensitivity for a given element. For actual samples, corrections for proton energy loss and matrix absorption of the X-rays must be included. These are comparable to similar corrections for absorption effects using photon excitation.

A number of studies have compared photon excitation with charged particle excitation for precision, accuracy, and minimum detectable limits.[34-36] Both methods are capable of analysis at concentrations of 1 ppm or greater and have an absolute particle limit of detectability in the range of 1 to 10 ng/cm^2. The accuracies and precisions are

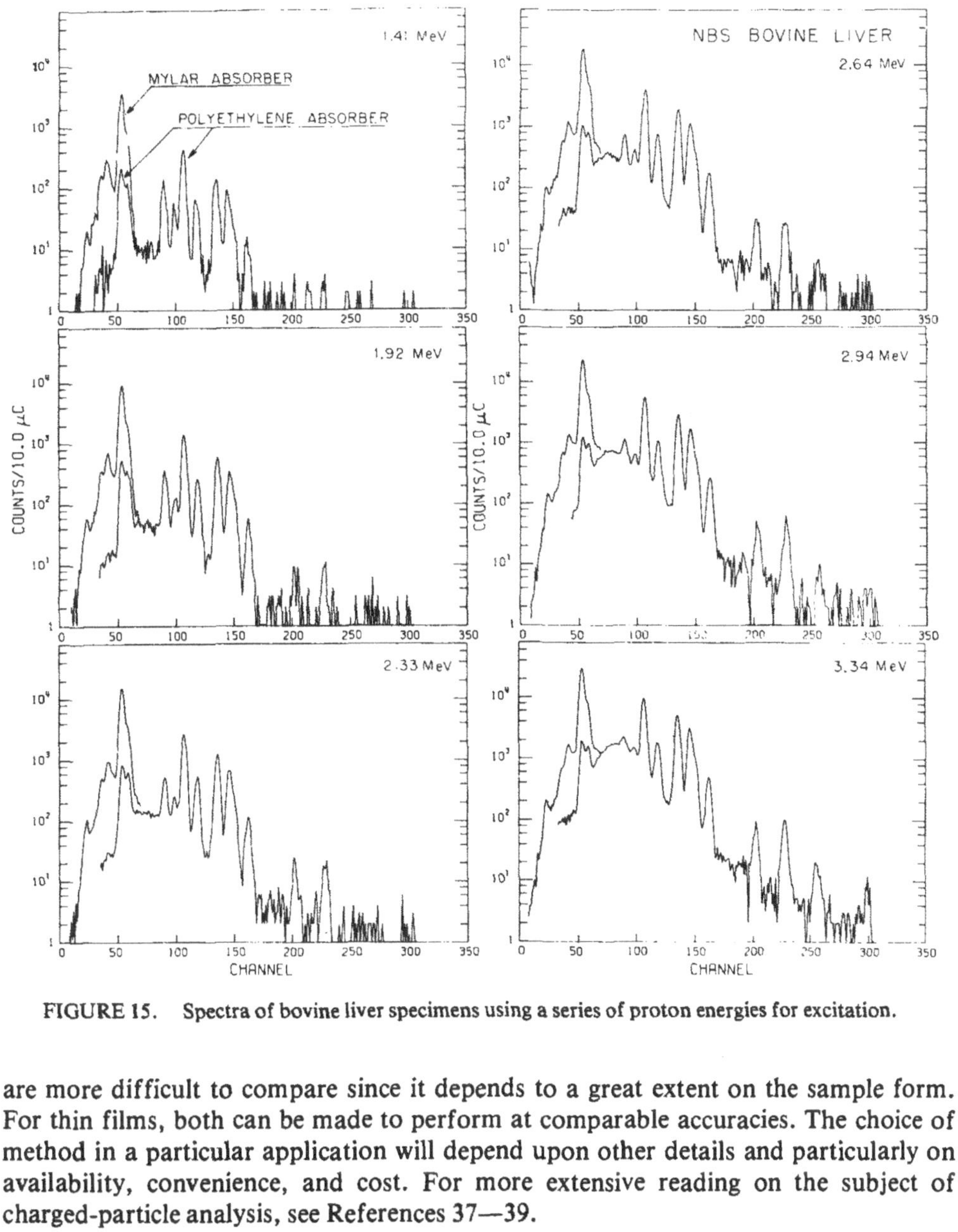

FIGURE 15. Spectra of bovine liver specimens using a series of proton energies for excitation.

are more difficult to compare since it depends to a great extent on the sample form. For thin films, both can be made to perform at comparable accuracies. The choice of method in a particular application will depend upon other details and particularly on availability, convenience, and cost. For more extensive reading on the subject of charged-particle analysis, see References 37—39.

V. ELECTRON BEAMS: MICROPROBES AND SCANNING ELECTRON MICROSCOPES

Direct excitation with electron beams is not routinely used on bulk specimens. The relatively poor sensitivity compared with other methods does not recommend its use. The principle application for electron excitation is in electron beam devices where the spatial resolution of finely focused beams is used to great advantage.

The two most commonly used electron beam devices are electron microprobes and scanning electron microscopes. Microprobes are characterized by higher beam currents (typically 10^{-6} to 10^{-9} A) and spatial resolutions of the order of 0.3 μm. They have been extensively used for quantitative analysis where the available beam currents make possible the use of wavelength dispersive spectrometry for analysis of elemental constituents.

Scanning electron microscopes (SEM) have significantly better spatial resolution (typically 0.01 μm (100 Å)) and operate at currents of 10^{-7} to 10^{-12} A. The normal mode of operation of one SEM utilizes a signal proportional to the intensity of secondary electrons emitted from the sample in order to generate a high resolution image of a portion of these samples. The electron beam is swept repeatedly across the specimen in a raster pattern while the secondary electron signal is synchronously displayed on a cathode ray screen.

The advent of semiconductor detectors with their inherently greater detection efficiency has permitted a simultaneous display of images extracted from individual characteristic X-ray intensities. It is thus possible to construct images of the elemental distributions of major constituents and relate them to the morphology of the specimen under study. The analysis of a particular area of the sample for several elements can be obtained by focusing the beam on a particular spot while a pulse height spectrum is accumulated. The sensitivity of such a measurement is limited by the contraints previously discussed; minimum detectable limits of less than 100 ppm for homogeneous samples are not practical.

The spatial resolution for X-ray images is not as good as can be obtained from secondary or backscattered electron images. X-ray signals can originate from anywhere within a volume surrounding the point of penetration of the electron beam, whereas backscatter images reflect only the surface layer where the beam enters.

Sample requirements are somewhat different for SEM and microprobe analysis. Since the most important information is related to the morphology of the particle, very little sample preparation is allowed. The sample is normally coated with a thin, conductive layer of carbon in order to prevent distortion due to charging on the sample. The specimen must be capable of withstanding the vacuum in the chamber without changing its composition.

High accuracies are normally not achieved in electron beam analysis particularly for samples such as biological specimens. The details of the electron beam loss together with the photon production and transport are not easily calculated. However, since most of the important information is contained in the relative spatial distributions of the elements, semiquantitative analysis is adequate in most cases.

Since the use of SEM techniques for the study of biological specimens is a highly developed technique for analysis, there are a number of review articles available for study. The reader is referred to the literature for more detailed information.[40,42]

VI. SPECIAL TOPICS

The previous sections dealt with variations in the application of XRF which are commonly employed and are likely to be available to the average user. There are several other variations of the method which are of additional interest either in special applications or because of their potential for future use. Several of these are discussed in the following section.

A. Proton Microprobe

Since charged particles beams can be focused, it is possible to combine the advantages of analytical sensitivity achieved in proton excitation with the spatial resolution available in microfocused beams. Several attempts to develop such a device have been made. Spatial resolutions of 2 μm have been achieved using a combination of fine collimators and magnetic focusing. Detection limits of a 1 to 10 ppm are possible for sample volumes of a few cubic micrometers. An interesting application of such a microprobe has been the measurement of the spatial distributions for several elements across the diameter of human hair.[19,43]

Although, photon excitation does not lend itself as readily to the production of fine focused beams, there have been attempts to develop small area analyzers using photon excitation. In one case, a crystal spectrometer is used to focus monochromatic radiation at a small line focus which then sees the sample.[44] In another method, a small collimator is placed on the detector in order to view an area of 1 mm diameter.[45]

B. Polarized X-Ray Sources

The background observed in the case of photon excitation is proportional to the intensity of the primary beam scattered in the sample. One method of reducing this intensity takes advantage of the fact that the amplitude for scattering is zero in the same direction as the electric vector of the incident electromagnetic radiation. Most sources of photons are unpolarized, i.e., their electric vectors are randomly oriented. However, it is possible by successive scattering of an unpolarized beam to generate a beam in which the vectors lie in a specific direction perpendicular to the direction of propagation. By placing the detector along this orientation, it is possible to minimize the amplitude of the scattered radiation. The photoelectric process, on the other hand, is insensitive to polarization and thus the fluorescence intensity is unaffected.

Several workers have demonstrated the expected reduction in background for a beam which had been partially polarized by successive scattering. The limitation on the widespread use of this technique is the inherent loss of intensity brought about the multiple scattering process.[5-7] Nevertheless, the technique has promise in the case of thick biological specimens.

C. Synchrotron Radiation

One method of obtaining intense, polarized beams is to use the radiation emitted by very high energy electron storage rings. There are a few facilities in the world where high-energy electron beam (3 to 5 GeV, i.e., 3×10^9eV) are circulated in large diameter ($\sim$100 m) storage rings. As the electrons undergo the acceleration required to maintain the circular orbit, synchrotron radiation is emitted tangentially to the orbit. This intense radiation is completely polarized, highly directional, and can extend to 30 to 40 keV in energy.

Using a crystal monochromator, it is possible to select a narrow energy band with which to excite samples. The polarization can be maintained while focusing a flux of 10^{11} photons/sec/mm^2 on the sample. This exceeds the capability of conventional X-ray tubes by a factor of 300. In one such series of measurements, a minimum detectable limit of less than 10^9 atoms in small geological samples was achieved. This correspond to 10^{-12} g of the element in question.[46,47]

D. Low Energy X-Ray Detection

The applications of XRF which have been discussed emphasize the detection of elements above Z — 11. This is due to the inability of X-rays from lighter elements to efficiently penetrate the 2.5×10^{-3} cm Be window normally used in Si(Li) spectrometers. It has been shown, however, that by incorporating the spectrometer into the same vacuum system as the sample, it is possible to observe elements down to and including boron, E_k = 185 eV.[48]

Figure 12 shows the detection of carbon X-rays using alpha-particle excitation. Similar spectra can be obtained using electrons or protons for excitation. Photon sources are normally not used for light element excitation because of the source attenuation which they undergo.

Although the configurations used for such measurements are inconvenient, they demonstrate a significant capability for light element analysis particularly when used in conjunction with microprobe and heavy ion excitation. Although the detection of very light elements is of dubious benefit in many biomedical applications, it is important to realize that Si(Li) spectrometers can be made capable of operation at these very low energies.

REFERENCES

1. **Dyer, G. R., Gedke, D. A., and Harris, T. R.,** Fluorescence analysis using an Si(Li) X-ray energy analysis system with low-power X-ray tubes and radioisotopes, *Adv. X-Ray Anal.*, 15, 228, 1972.
2. **Jaklevic, J. M., Giauque, R. P., Malone, D. F., and Searles, W. L.,** Small X-ray tubes for energy dispersive analysis using semiconductor spectrometer, *Adv. X-Ray Anal.*, 15, 266, 1972.
3. **Zeigler, C. A., Ed.,** *Applications of Low Energy X- and Gamma-Rays,* Gordon and Breach Science Publishers, New York, 1971.
4. **Rhodes, J. R.,** Design and application of X-ray emission analyzers using radioisotope X-ray or gamma-ray sources, *Energy Dispersion X-Ray Analysis,* ASTM Special Technical Publication 485, American Society for Testing and Materials, Philadelphia, 1971, 243.
5. **Kaufman, L. and Camp, D. C.,** Polarized radiation for X-ray fluorescence analysis, in *Adv., X-Ray Anal.*, 18, 247, 1975.
6. **Howell, R. H., Pickles, W. L., and Cafe, J. L.,** X-ray fluorescence experiments with polarized X-rays, in *Adv. X-Ray Anal.*, 18, 265, 1975.
7. **Dzubay, T. G., Jarrett, B. V., and Jaklevic, J. M.,** Background reduction in X-ray fluorescence spectra using polarization, *Nucl. Instrum. Methods,* 115, 297, 1974.
8. **Lifshin, E.,** Solid-state X-ray detectors for electron microprobe analysis, *Energy Dispersive X-Ray Analysis; X-Ray and Electron Probe Analysis,* ASTM STP 485, American Society for Testing and Materials, Philadelphia, 1971, 141.
9. **Russ, J. C.,** Energy dispersive X-ray analysis on the scanning electron microscope, in *Energy Dispersive X-Ray Analysis; X-Ray and Electron Probe Analysis,* ASTM STP 485, American Society for Testing Materials, Philadelphia, 1971, 154.
10. Proc. int. conf. particle induced X-ray emission and its applications, *Nucl. Instrum. Methods,* 142, 1977.
11. Proc. 3rd Conf. Application of Small Accelerators, ERDA Report CONF-741040, Duggan, J. L., North Texas State University, October 21, 1974.
12. **McMaster, W. H., Kerr del Grande, N., Mallett, J. H., and Hubbell, J. H.,** Compilation of X-ray Cross Sections, UCRL-50174, Lawrence Livermore Laboratory, 1970.
13. **Jaklevic, J. M., Goulding, F. S., Jarrett, B. V., and Meng, J. D.,** Application of X-ray fluorescence techniques to measure elemental composition of particles in the atmosphere, *Analytical Methods Applied to Air Pollution Measurements,* Stevens, R. K. and Herget, W. F., Eds., Ann Arbor Science Publishers, Ann Arbor, Mich., 1974, 123.
14. **Gedke, D. A., Elad, E., and Denee, P. B.,** An intercomparison of trace element excitation methods for energy-dispersive fluorescence analyzers, *X-ray Spectrom.*, 6, 21, 1977.
15. **Powell, C. J.,** Cross sections for ionization of inner-shell electrons by electrons, *Rev. Mod. Phys.*, 48, 33, 1976.
16. **Green, M. and Cosslett, V. E.,** The efficiency of production of characteristic X-radiation in thick targets of a pure element, *Proc. Phys. Soc. London,* 78, 1206, 1961.
17. **Jaklevic, J. M.,** Excitation Methods for Energy Dispersive Analysis, presented at Proceedings ERDA X- and Gamma-Ray Symposium, Ann Arbor, 1976.
18. **Garcia, J. D.,** X-ray production cross sections, *Phys. Rev. A,* 1, 1402, 1970.
19. **Nobiling, R., Traxel, K., Bosch, F., Civelekoglu, Y., Martin, B., Poul, B., and Schwalm, D.,** Focusing of proton beams to micrometer dimensions, *Nucl. Instrum. Methods,* 142, 49, 1977.
20. **Giauque, R. D., Goulding, F. S., Jaklevic, J. M., and Pehl, R. H.,** Trace element analysis with semiconductor detector X-ray spectrometers, *Anal. Chem.*, 45, 671, 1973.
21. **Jenkins, R.,** *An Introduction to X-ray Spectrometry,* Heyden & Son, New York, 1974.
22. **Birks, L. S.,** *X-ray Spectrochemical Analysis,* 2nd ed., Interscience Publishers, New York, 1969.
23. **Leibhafsky, H. A., Pfeiffer, H. G., Winslow, E. H., and Zemany, P. D.,** *X-ray, Electrons, and Analytical Chemistry,* Wiley-Interscience, New York, 1972.
24. **Camp, D. C., Cooper, J. A., and Rhodes, J. R.,** X-ray fluorescence analysis — results of a first round intercomparison study, *X-ray Spectrom.*, 3, 47, 1974.

25. Camp, D. C., van Lehn, A. L., Rhodes, J. R., and Pradzynski, A. H., Intercomparison of trace element determinations in simulated and read air particulate samples, *X-ray Spectrom.*, 4, 123, 1975.
26. Alfrey, A. C., Nunnelley, L. L., Rudolph, H., and Smythe, W. R., Medical applications of a small sample X-ray fluorescence system, *Adv. X-ray Anal.*, 19, 497, 1976.
27. Cooper, J. W., Comparison of particle and photon excited X-ray fluorescence of environmental samples, *Nucl. Instrum. Methods*, 106, 525, 1973.
28. Musket, R. G., private communication.
29. Goulding, F. S. and Jaklevic, J. M., Photon-excited energy-dispersive X-ray fluorescence analysis for trace elements, *Annu. Rev. Nucl. Sci.*, 23, 45, 1973.
30. Jaklevic, J. M. and Goulding, F. S., Energy dispersion, *X-ray Spectrometry*, Herglotz, H. K. and Birks, L. S., Eds., Marcel Dekker, New York, 1978.
31. Woldseth, R., *X-ray Energy Spectrometry*, KeVex Corporation, Burlingame, California, 1973.
32. Willis, R. D. and Walter, R. L., private communication, 1978.
33. Willis, R. D., Walter, R. L., Shaw, R. W., and Gutknetht, W. F., Proton-induced X-ray emission analysis of thick and thin targets, *Nucl. Instrum. Methods*, 142, 67, 1977.
34. Jaklevic, J. M. and Walter, R. L., Comparison of minimum detectable limits among X-ray spectrometers, *X-ray Fluorescence Analysis at Environmental Samples*, Dzubay, T. G., Ed., Ann Arbor Science, Ann Arbor, 1978, 63.
35. Goulding, F. S. and Jaklevic, J. M., XRF Analysis — Some sensitivity comparisons between charged-particle and photon excitation, *Nucl. Instrum. Methods*, 142, 323, 1977.
36. Ahlberg, M. S. and Adams, F. S., Experimental comparison of photon- and particle-induced X-ray emission analysis of air particulate matter, *X-ray Spectrom.*, 7, 73, 1978.
37. Nelson, J. W., Proton-induced aerosol analysis: methods and samplers, *X-ray Fluorescence Analysis of Environmental Samples*, Dzubay, T., Ed., Ann Arbor Science, Ann Arbor, 1977, chap. 2.
38. Johansson, T. B., Akselsson, and Johansson, S. A. E., Proton-induced X-ray emission spectroscopy in elemental trace analysis, *Adv. X-ray Anal.*, 15, 373, 1972.
39. Nielson, K. K., Hill, M. W., and Mangelson, N. F., Calibration and correction methods for quantitative proton-induced X-ray emission analysis of autopsy tissues, *Adv. X-ray Anal.*, 19, 511.
40. Tousimis, A. J. and Marton, L., Eds., *Advances in Electronics and Electron Physics*, Vol. 6, Academic Press, New York, 1969.
41. Birks, L. S., *Electron Probe Microanalysis*, Interscience, New York, 1963.
42. Heinrich, K. F. L., Ed., *Quantitative Electron Probe Microanalysis*, NBS Publication, No. 298, National Bureau of Standards, U.S. Department of Commerce.
43. Cookson, J. A. and Pilling, F. D., Trace element distributions across the diameter of human hair, *Phys. Med. Biol.*, 20, 1015, 1975.
44. Ong, P. S. and Cox, Jr., H. L., Line-focusing X-ray monochromator for the analysis of trace elements in biological specimens, *Med. Phys.*, 3, 74, 1976.
45. Jaklevic, J. M., French, W. R., Clarkson, T. W., and Greenwood, M. R., X-ray fluorescence analysis applied to small samples, *Adv. X-ray Anal.*, 21, 171, 1978.
46. Sparks, C. J., Raman, S., Yakel, H. L., Gentry, R. V., and Krause, M. O., Search with synchrotron radiation for superheavy elements in giant-halo inclusions, *Phys. Rev. Lett.*, 38, 205, 1977.
47. Sparks, C. J., Ricci, E., Raman, S., Krause, M. O., Yakel, H. L., and Gentry, R. V., X-ray Fluorescence Analysis and Synchrotron Radiation, *Fourth Annual Stanford Synchrotron Radiation Laboratory Users Group Meeting*, SSRL Report No. 77/11, Stanford Synchrotron Radiation Laboratory, Stanford, 1977, 58.
48. Musket, R. G., Detection of proton-induced boron X-rays with a Si(Li) detector, *Nucl. Instrum. Methods*, 117, 385, 1974.

Section II
Trace Elements

Chapter 4

TRACE ELEMENTS IN MEDICINE

P. S. Ong

TABLE OF CONTENTS

I. Trace Elements in Man 71
A. Introduction 71
B. Trace Elements in Serum 73
C. Trace Elements in Tissue 74

II. Fluorescence Radiation-Induced Energy-Dispersive Analyzer (FRIEDA) 74
A. Introduction 74
B. The FRIEDA System 75
C. The X-Ray Probe 76
D. Performance of the FRIEDA System 77

III. Applications of the FRIEDA System 79
A. The Trace Element Profile in Serum 79
B. Trace Elements in Hair 80
C. Metalloproteins 81

References 87

I. TRACE ELEMENTS IN MAN

A. Introduction

A comparison of the relative abundance of the chemical elements in human serum[1] with that of the earth's crust[2,3] reveals striking similarities in composition. Even more striking is the similarity between serum and seawater[4] (Figure 1). Through the evolutionary process, certain elements that are abundant in seawater have become selectively used for certain biological functions. The elements most abundantly available were used as major building blocks. The elements present in traces were more suitable for future complex metabolic functions. These trace elements were probably initially used as catalysts to perform simple chemical reactions, or acted as links to create complex and active protein molecules. Elements originally found suitable in the performance of certain functions might, however, have fallen behind in the evolutionary process. Examples of such poor choices are the use of vanadium in the blood cells of ascidians, and the use of copper instead of iron in the blood of worms, crustacea, mollusks, and insects.[5]

When man emerged some 2 million years ago, he became a passive member of the naturally stable, but dynamically evolving, ecosystem.[6] Changes in the environment were minor, occurring over long periods of time. Contemporary man, however, has introduced new contaminants which have disrupted the natural stability in areas where the highest population densities are concentrated. Leaded gasoline used by cars introduces 150,000 tons of lead into the air annually in this country alone;[7] the lead content in air increases or decreases daily with traffic density[8] and with distance from high-

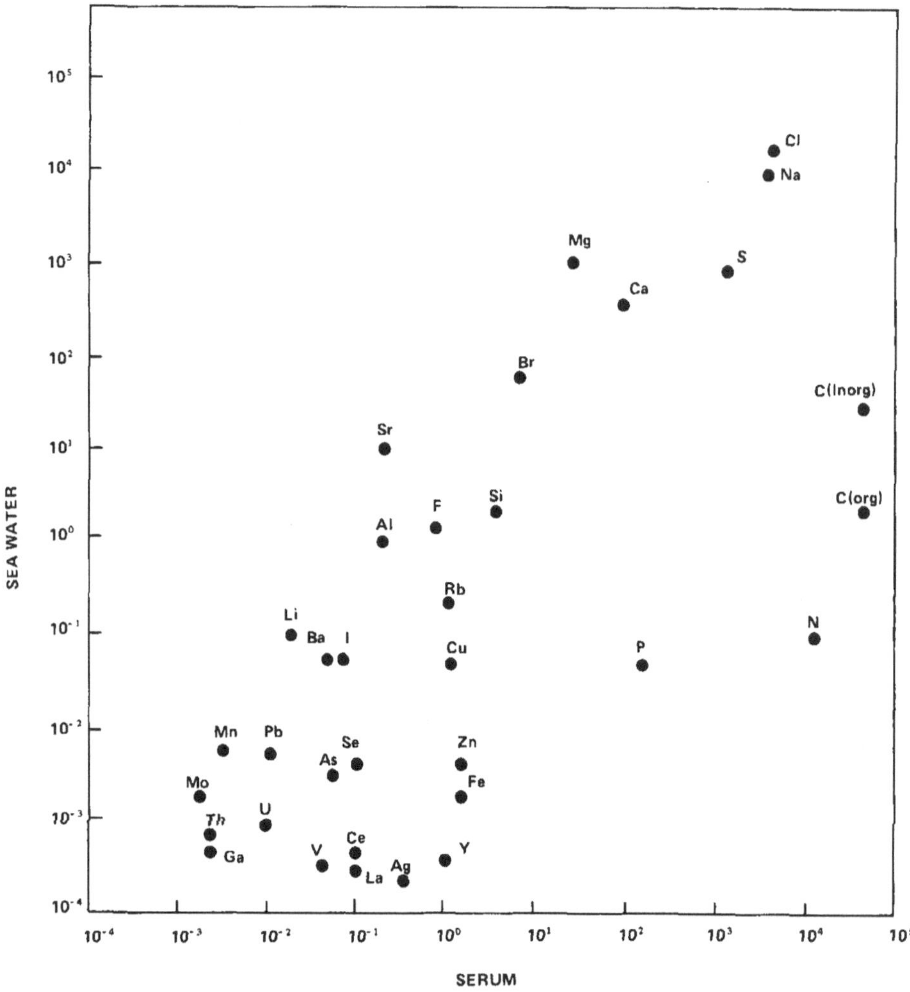

FIGURE 1. Trace element comparison between seawater and normal human serum. The scales are in parts per million. (Adapted from Bowen[1] and Sverdrop et al.[4])

ways. Industry is polluting the air with cadmium, mercury, arsenic,[7] and other elements which were not normally present in such concentrations during evolutionary periods. Thus, modern man has introduced many new sources for trace element toxicity.

Presently, there are ten trace elements which are recognized to perform functions essential to the life or health of man. These are chromium, manganese, iron, cobalt, copper, zinc, selenium, molybdenum, tin, and iodine. Man has not, however, inherited a high degree of tolerance to certain elements, usually classified as toxic and among which are those now found in the air — beryllium, nickel, cadmium, lead, selenium, mercury, arsenic, and tin.[9]

While trace element deficiency can be understood easily in terms of a lack of building materials for the production of life's essential compounds, the subject of toxicity or poisoning is quite complex. It may relate to a displacement of essential elements from metabolically active sites by toxic atoms which are chemically similar, thus manifesting itself as a deficiency in the element being replaced. Or, it may act directly as a cellular toxin by changing the physical properties of the proteins which result in denaturization.

Elements with the same outer-shell electron configuration, that is, belonging to the same group in the periodic table, tend to substitute each other; thus, magnesium could be replaced by beryllium; potassium by rubidium, cesium, or thallium; sulphur by tellurium or selenium; and zinc by cadmium or mercury. These substitutions may, in rare cases, stimulate or alter functions of an enzyme. In most cases, a substitution depresses such functions.[5]

B. Trace Elements in Serum

Serum is one of the most suitable materials for the study of trace elements in man. It supplies all parts of the human body with their necessary metabolic components (except oxygen). It is readily available and can be obtained without undue discomfort or risk to the patient.

The importance of serum trace metals in human physiology was recognized as early as 1857.[10] Since then, numerous specific diseases have been shown to be associated with altered concentrations of certain trace elements. Copper is one of the many trace metals in serum which has been the subject of extensive investigation, primarily because a reasonably consistent normal value could be defined, and also because striking changes in serum copper levels (SCL) have been observed in a variety of diseases.

Warburg and Krebs[11] reported in 1937 an increased SCL in pregnancy. Locke et al.[12] in 1932 tabulated several diseases causing SCL to increase — leukemia, pernicious anemia, Addison's disease, diabetes, hyperthroidism, carcinoma of various types, and various kinds of liver disease. Later, Sachs et al.[13] contributed additional disorders to the list — iron deficiency anemia, sickle cell anemia, polycythemia vera, purpura, Banti's syndrome, the collagen diseases, malaria, arsenic poisoning, schizophrenia, and various disorders of the nervous system.

Since that time, reports of diseases related to hypercupremia have increased. Heilmeyer et al.[14] included aplastic anemia and allergic reactions, and Cartwright[15] added Hodgkin's disease. Several reports by Pagliardi,[16] Jensen,[17] Cherry,[18] and others[19-21] have indicated correlations between serum copper levels in both stage of disease and response to therapy, raising hopes that serum copper level might be a useful simple technique to monitor effectiveness of therapy and maintenance of remission. This has not come into wide general use.

Only a few reports of disease-related hypocupremia have appeared — hepatolenticular degeneration (Wilson's disease) by Cummings,[22] Menke's Kinky Hair disease by Craver,[23] and lipoid nephrosis and dysproteinemia of infancy by Adelstein et al.[24]

Less dramatic, but equally important, are the published observations on zinc, iron, and nickel in human serum. Decreased serum zinc levels have been observed in myocardial infarction, lymphoma, carcinoma, pneumonia, and aplastic anemia.[25] Abnormal concentrations of nickel in serum have been reported in patients with myocardial infarction, stroke, burns, hepatic cirrhosis, and uremia.[26] The serum iron level is a common indicator of bone marrow red blood cell production. Craver[23] has reported low serum iron level as one of the first signs of activity in Hodgkin's disease, although serum iron may be elevated in the presence of hemolytic anemia or suppressed crythropoiesis. Jaffee et al.[27] found hypoferremia to occur in relapse and in many instances of remission in advanced stages of Hodgkin's disease. Markedly depressed and markedly elevated serum iron levels will be seen in severe iron deficiency and in iron overload states (e.g., hemochromatosis), respectively.

A change in one trace element is often accompanied by an inverse change in the level of another trace element. Decreased serum iron associated with increased serum copper has been described by Brenstrup,[28] Cartwright et al.,[15] and Elvehjem.[29] The combination of zinc and copper was studied by Sinha et al.,[30] who also reported a decrease in zinc levels in alcoholism, Laennec cirrhosis, bronchitis, and pneumonia.

Disease-related trace element levels have general value only if a normal range suitable for comparison can be defined. Sinha et al.[30] have added their data to a review of various authors between 1957 and 1967 to characterize the normal levels and ranges of serum copper and zinc. Bowen[1] compiled a comprehensive list of normal values for all trace elements in mammalian blood and serum, but the normal levels of various other specific elements have been reported only sporadically.[31,32]

C. Trace Elements in Tissue

In the past, various investigators have measured the composition of human tissues taken from a variety of organs in order to define the composition of Standard Man.[33-35] Attempts to correlate tissue composition with occurrence of cancer has so far not proven to be either useful or conclusive, even though compositional differences have been observed. For example, Danielson and Steinnes[36] found that the zinc-to-potassium ratio of cancerous tissue is significantly lower than that in normal tissue. Mulay et al.[37] who compared 22 trace elements in cancerous and noncancerous tissue taken from various organs, found significant differences pertain only to a specific type of cancer. Janes et al.[38] investigated normal bone and human osteogenic sarcoma for eight trace elements. They found an increase in iron, zinc, copper, and manganese, while calcium, magnesium, cobalt, and nickel were decreased.

Of all the body's tissues, hair is probably the most easily available for study, raising hope for its use as a uniformly applicable, noninvasive biopsy material.[39] Its most positive application would appear to be in the detection of metal poisoning. Lead, mercury, and arsenic poisoning are readily detectable by the presence of large amounts of these elements in hair. On the other hand, trace metal deficiencies can be diagnosed effectively by reduced levels in hair, such as zinc deficiency.[40]

Since hair grows at a rather constant rate, and important elements are tightly bound in the patient's hair, the distribution of the elements along the length of a hair makes it possible to estimate the time of exposure.[41] It is extremely important to realize, however, that there are many trace elements present in the air, in hair shampoos, and in other chemical hair preparations which are extremely difficult to remove even by repeated washings, and which can be easily misinterpreted as elements reflecting increased systemic exposure.[43]

It has been known for some time that patients with Wilson's disease have increased deposition of copper in the liver, brain and kidney. A study by Rice and Goldstein,[44] however, showed that the copper content of nails, hair, and skin of these patients is normal.

II. FLUORESCENCE RADIATION-INDUCED ENERGY-DISPERSIVE ANALYZER (FRIEDA)

A. Introduction

Fluorescent excitation analysis exhibits many of the most desirable features for multi-elemental analysis of biological samples. With the exception of the very light elements, the relative sensitivity is adequate for the trace elements of biological interest. Interelement interference, one of the major limitations in the energy dispersive system, is with exception of a few cases, not very serious because of the natural trace elements pattern in biological samples. The range of relative concentrations in which the important elements are present in body fluids and tissue is wide, covering several orders of magnitude, but it is easy to measure these elements simultaneously because the sensitivity curve of the instrument varies inversely with the relative abundance.

Because the technique is nondestructive, the sample remains essentially unchanged after analysis and may be used for further study with the same technique or any other technique.

B. The FRIEDA System

The FRIEDA system is based on the use of monochromatic radiation which is obtained with a secondary fluorescent source.[45] One advantage of such an approach is that monochromatic radiation at the desired energy may be selected without changing the primary X-ray tube source. FRIEDA was built around a Philips manual vacuum spectrograph, which employs a sealed-off X-ray tube. Sample cups from the original unit are used to hold the pure elements which act as secondary radiators. The instrument can house four different secondary fluorescent sources which are interchangeable without breaking the vacuum. The original goniometer has been replaced by a new sample chamber, and the collimator has been modified by a simple baffle. This was necessary in order to screen off radiation coming from the periphery of the radiator. Figure 2 shows schematically the geometrical arrangement. The excitation radiation intersects the sample at 45° and the emergence angle is 45°. An Mo tube, operated at 45 kV and 30 mÅ, is used in most instances. With an Se radiator (energy 15.30 KeV), serum Fe, Cu, and Zn are analyzed in about 20 min.

The signal from the detector is fed into a 1024 channel pulse height analyzer (Northern NS 636). This analyzer is interfaced to a minicomputer (Xerox CF-16) and controlled by the operator by means of a teletype. The spectra are recorded on magnetic tape (Houston Scientific), plotted by an X-Y recorder (Complot), or printed on paper (Teletype).

The major problem of the FRIEDA system has been in obtaining a radiator material of adequate purity. Selenium was initially used, but it is difficult to obtain free of copper. Even the selenium which is specified as 99.999% pure (A. D. MacKay, Inc.) shows a copper peak equivalent to 10% of the measured copper concentration of a normal serum sample. In subsequent work, Zr, which is available in higher purity, was used as a secondary radiator (Figure 3).

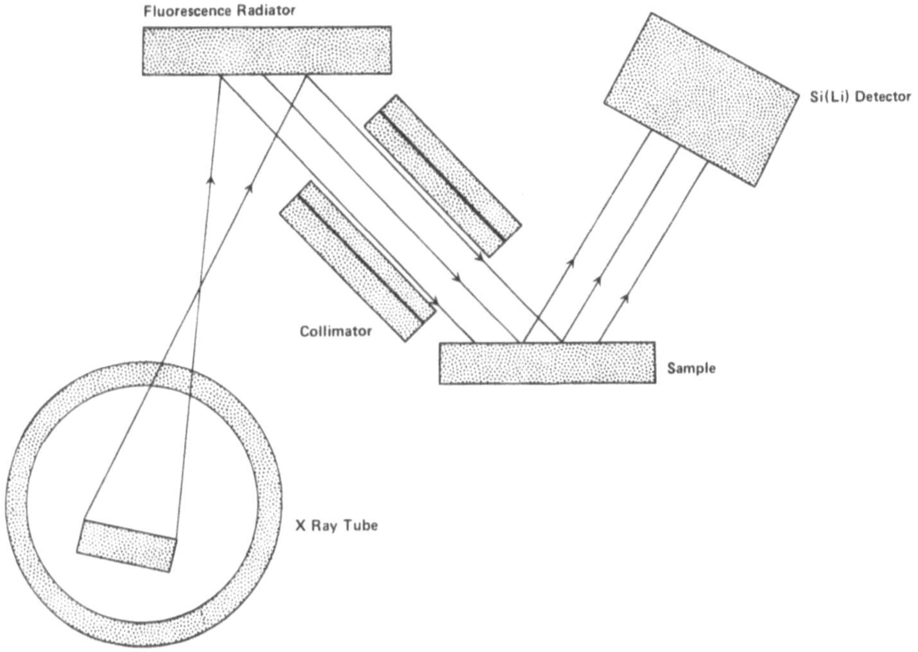

FIGURE 2. The geometry of the X-ray tube, radiator, sample, and detector of the FRIEDA System.

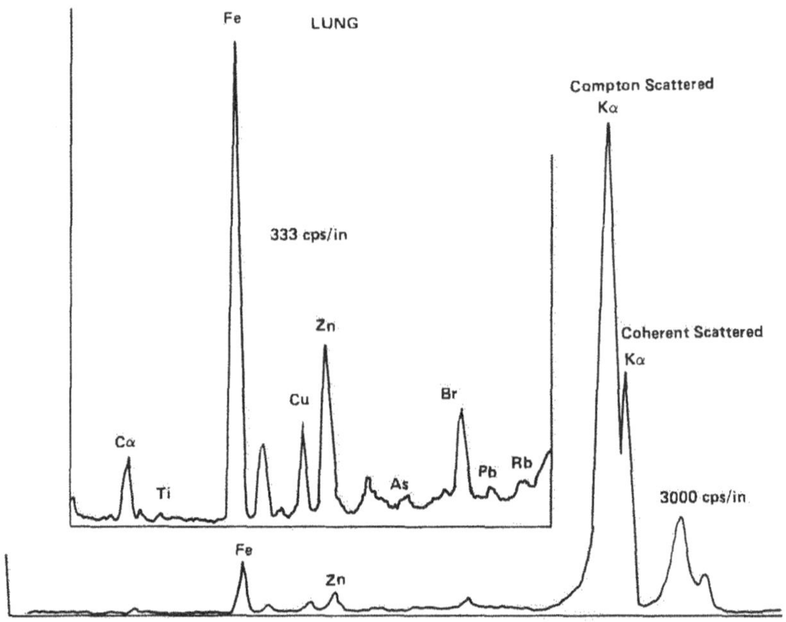

FIGURE 3. Trace element using the Zr Kα and Kβ lines for excitation. The insert is plotted on a ten-times-expanded scale.

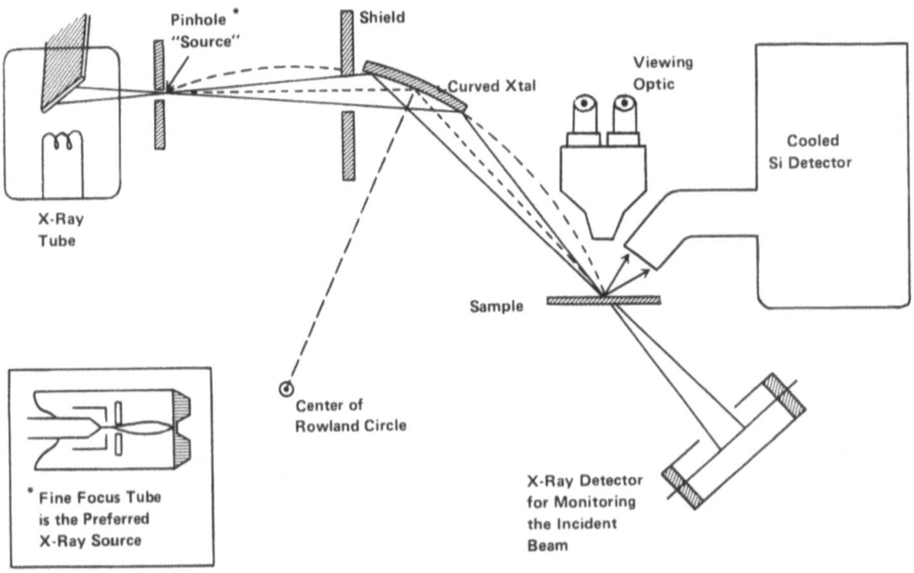

FIGURE 4. *Schematic drawing of the focused X-ray probe.*

C. The X-Ray Probe

The use of a focusing crystal results in an X-ray spot of high intensity. The sample to be analyzed, therefore, can be reduced to the size of the focus dimension without loss of fluorescence sensitivity.

Figure 4 shows the layout of the focusing monochromator. It consists of a standard X-ray tube, a small apeture which serves to create a pinhole source, a scatter shield, the diffraction curved crystal, a viewing optic, a sample holder, and the energy dispersive detector. In the present arrangement, an Mo tube is being used with a $K\alpha = 17.479$

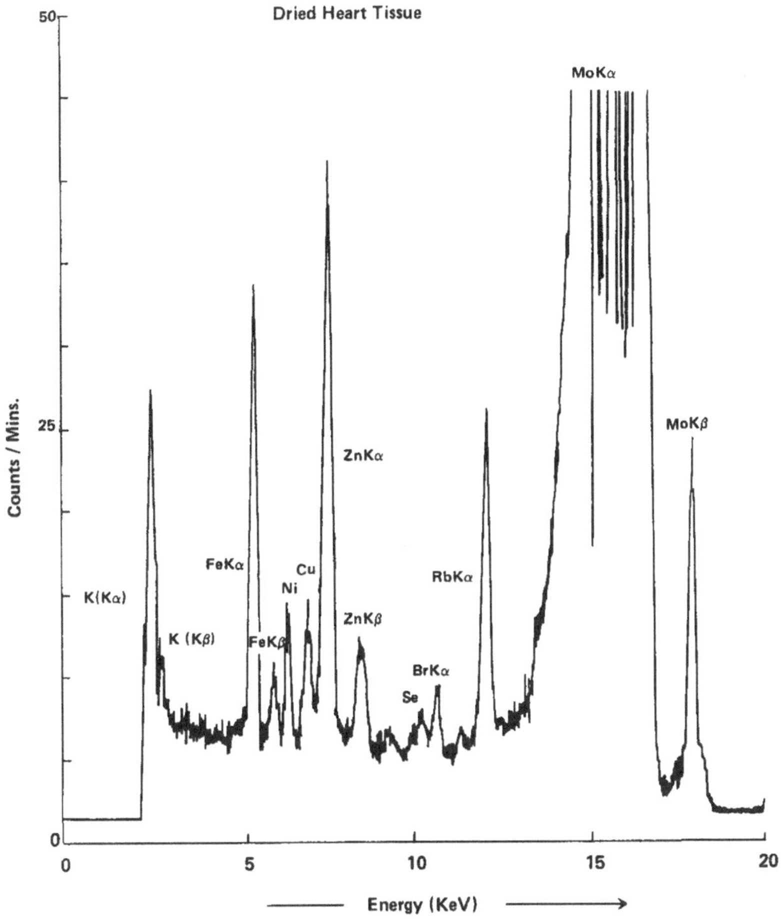

FIGURE 5. Spectrum of dried heart tissue excited with the focused X-ray probe.

KeV. The pinhole is 0.5 mm in diameter and is made of lead. All shields are lead, lined with a sheet of Mo. The crystal is curved in one direction only, to a radius of curvature of 209.6 mm. The crystal is mounted on a linear stage to accommodate various distances, and on a gimbal to permit rotation for selection of the Bragg angle. At present, the sample stage and detector arrangement are improvized, and no optical system is included.

The dimension of the line focused image of the monochromatic beam is 0.5 × 40 mm. The intensity at the focal plane, measured with a proportional counter, is in excess of 10^7 c/s for 40 KV and 30 mA. An example of a spectrum obtained with this instrument is shown in Figure 5. The use of the zirconium Kα X-ray as an exciting source with an energy of 15.7751 KeV is more effective. This lower energy radiation will produce less Compton scattering and is more efficient for the excitation of all the elements from Br (Z = 35) down. A more detailed description is given by Ong and Cox.[46]

D. Performance of the FRIEDA System

The precision of the FRIEDA system has been tested for a large number of pooled sera, in which Ni was added as an internal standard. The intensities from the trace elements Fe, Cu, and Zn are determined and compared with that on Ni. The same sample was measured ten times and yielded a relative standard deviation of 0.7% for Cu to Ni; and of 0.3% for Zn to Ni. These were obtained for Cu to Ni and Zn to Ni ratios of 0.78 and 0.56 respectively. The element Ni actually is not a good choice for use as an internal standard for serum, since normal serum contains about 0.1 ppm Ni, which may vary among individuals. The element Ge would be a better standard.

Since an internal standard is used, the precision as determined corresponds to the statistical variation in the measured total number of characteristic X-ray photons and on background. Intensity variations of the X-ray tube are eliminated because ratios are taken. It may be necessary to use the instrument without an internal standard. One must then determine the intensity variation which occurs during the course of an average day, as well as from day to day. This is done using the integrated Compton peak, since its inherent statistical variation is small. The results show that the variation which occurs during a 12-hr period is 1.6%. This variation is due to a constant drift in the X-ray output. Corrections can be made to reduce this to negligible values. This day-to-day intensity variation, which is caused by the resetting of current and voltage to the X-ray tube, requires that an intensity calibration be performed daily.

Other factors which affect the analytical precision include the sample preparation technique, the use of internal standards and the variation in protein content of the serum resulting from variations in sample thickness. To determine these factors for serum analysis, about 50 samples were made from pooled serum, each treated as a separate sample and prepared during the course of several days. Each sample consisted of 2 mℓ serum measured with a disposable plastic pipette. Using an Oxford pipette with disposable tip, 0.200 mℓ of 20 ppm $NiCl_2$ solution was added and ultrasonically mixed.

The technique as originally used requires that the sample be ashed in a low temperature asher (Trapelo). Such an additional operation increases the danger of introducing new errors. Since the absorption of Fe radiation by the protein matrix cannot be ignored if the sample is not ashed, an absorption correction must be applied. A correction procedure has been developed based on the use of two elements as internal standards: one element with a characteristic X-ray line for which the matrix absorption is small; the other one with a radiation heavily absorbed by the matrix. In essence, the former characterizes the unabsorbed quantity of that element, while the variation of the measured intensity of the latter indicates the degree of absorption.

The sensitivity of the FRIEDA system has been determined using standards prepared by Columbia Scientific Instruments in Austin, Texas. These consist of Whatman 41 or millipore filter papers on which known amounts of solutions of various single elements have been uniformly deposited. Since the beam size of FRIEDA can be accurately determined (1.5 cm^2), the sensitivity can be expressed directly on an absolute scale as shown in Figures 6 and 7 and Table I.

The accuracy, i.e., the deviation of the analytical results from the true value, depends on the accuracy of the values given by the standards, and on the uncertainties due to interelement effects and self absorption. In biological samples, interelement effects are negligibly small for the trace elements. Self-absorption, however, can be considerably higher if the sample is relatively thick. In serum samples, corrections can be applied as previously discussed.

The sample mass is determined by quantitating the Compton scattered radiation from the matrix. For a standard, either pure carbon or a protein sample can be used. The difference in scatter factor between the real sample and the standard introduces an error which can be corrected, as described by Cox and Ong[47] for an average protein sample. The results on liver tissue samples indicate that the thus-corrected values of Fe (294 ppm), Cu (199 ppm), and Zn (199 ppm) fall within the certified values of the NBS standard.[47]

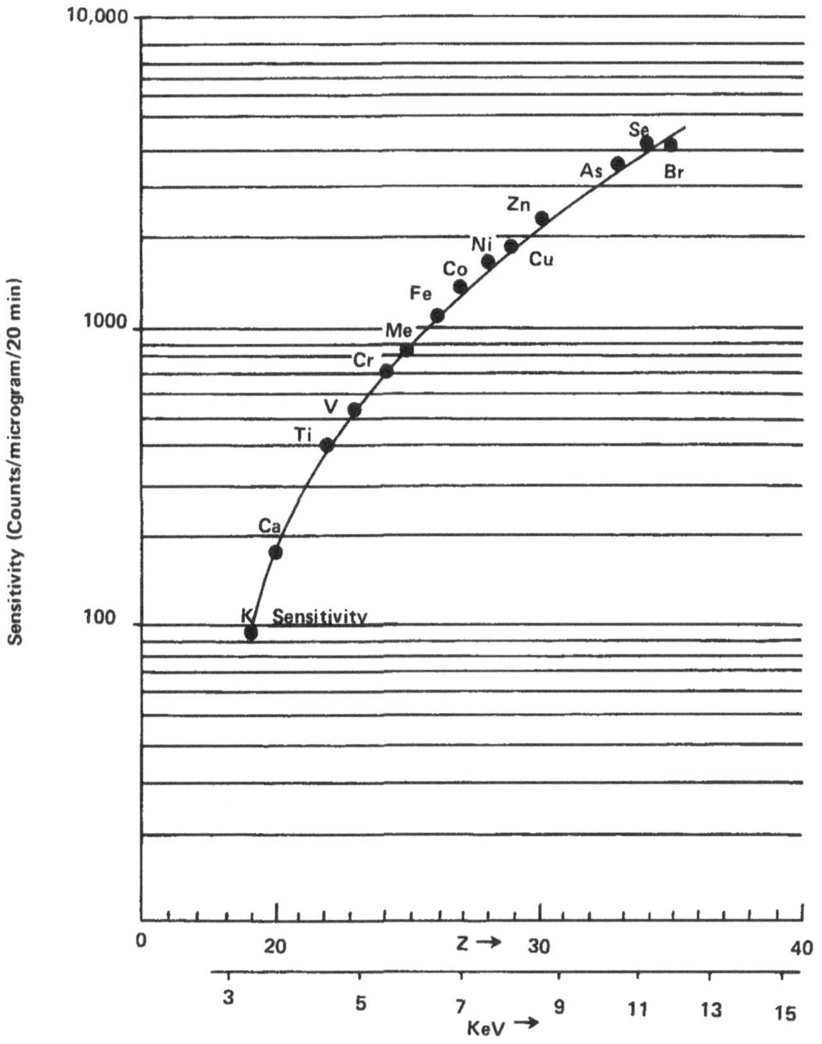

FIGURE 6. The count sensitivity of the FRIEDA system for various elements obtained in a counting time of 20 min.

III. APPLICATIONS OF THE FRIEDA SYSTEM

A. The Trace Element Profile in Serum

Figures 9 and 10 illustrate serum profiles using the FRIEDA system and covering three trace elements and five major elements in two families carrying the disease, Menke's Kinky Hair Syndrome (MKHS), and also shows a profile from a normal subject for comparison. Patients with MKHS have a genetic defect in copper metabolism. As in Wilson's disease, serum copper levels are low, but in contrast, patients with MKHS have very low levels in tissues, including hair. The patient usually does not survive to adulthood. The serum trace element profile of WM (Figure 8) shows a negligible amount of copper in comparison with that of a normal donor. It is remarkable that the patient's mother (M) as well as his father (F) show an increase of serum zinc as compared to that of the normal.

A similar observation has been made in the parents and a sister of another patient, as shown in Figure 9. Note also the high copper level of the mother (M) because of

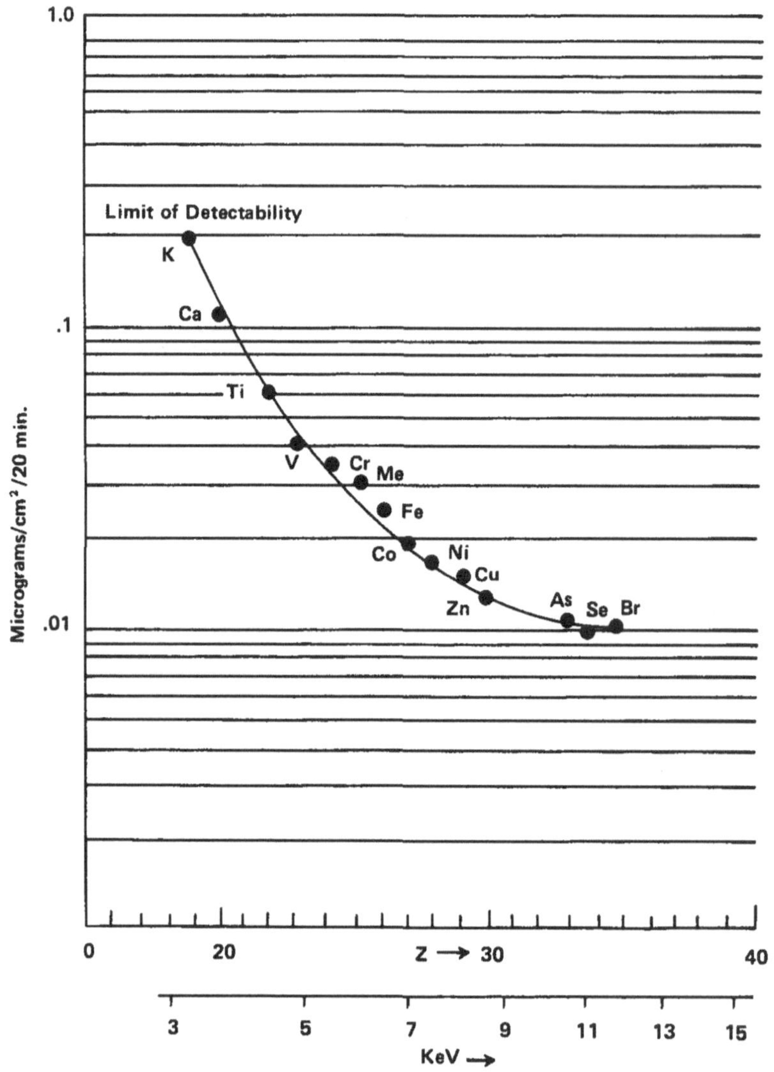

FIGURE 7. Detection limit of the FRIEDA system in μg/cm² (3 standard deviation above background in 20 min) using Zr for excitation.

birth control pills. The limited number of observations (2) does not permit any conclusion to be made. (The serum samples were kindly supplied by Dr. R. Zeller, Methodist Hospital, Houston.)

B. Trace Elements in Hair

One can be certain that before changes in the basal profile reflect a specific disease, a trace element profile should be established for each individual in a healthy condition. When a trace element profile is taken at a time when symptoms are present, there is no established method for determining the patient's basal profile prior to these symptoms. Since hair grows continuously (about 10 mm/month for head hair), most persons have hair growth covering at least a 3-month period. The hair cells, before keratinization, were part of the living tissue and were nourished by blood. While other cells are eventually broken down and the nonessential or trace elements are disposed by excretion, hair cells maintain and preserve a permanent record along the length of a hair. Therefore, the body's trace element history may still be found in hair growth.

TABLE I

Performance of the FRIEDA System

Zr Excitation, 20 min Counting Time

	Detection limits[a]			Sensitivity
Elements	(μg)	(μg/cm²)	(μg/g)	(c/μg/cm²)
K	1.4	0.9	100	142
Ca	0.17	0.11	12	285
Ti	0.09	0.06	6.7	600
V	0.06	0.04	4.4	765
Cr	0.052	0.035	3.9	1050
Mn	0.095	0.030	3.3	1275
Fe	0.038	0.025	2.8	1650
Co	0.030	0.020	2.2	1950
Ni	0.025	0.017	1.9	2550
Cu	0.019	0.013	1.4	2850
Zn	0.018	0.012	1.3	3300
As	0.015	0.010	1.1	3600
Se	0.015	0.010	1.1	6150
Br	0.015	0.010	1.1	6150
Cd(Lα)	1.0[b]	0.60[b]	66[b]	52
Pb(Lα)	0.03[c]	0.020[c]	2.1[c]	1828

[a] 3 standard deviations above background.
[b] Calculated from Ca background.
[c] Calculated from As background.

For example, zinc deficiency can be diagnosed through the lack of this element in hair.[45] In addition, lead and arsenic poisoning are reflected in an increase of these elements in hair.

This is illustrated in Figure 10. Figures 10a and 10b show two trace element spectra of hair samples from a girl (L) and her father (P). The presence of lead in the girl's hair is apparent. The distribution of various elements plotted on an arbitrary time scale is shown in Figure 11.

C. Metalloproteins

Although the significance of trace metals in man is recognized, the progress in studying their metabolic function has thus far been slow, and the clinical usefulness limited. Total trace metal determination in biomedical samples is considered as an initial step only for identification of a problem. The next step would be determination and investigation of the proteins associated with those metals, the metalloproteins. Among these metalloproteins are the important metalloenzymes.

Since many enzymes contain metal atoms, even if only a few per molecule, these metals can serve as an indicator of the presence of a specific metalloenzyme. Previous studies on metalloproteins have been aimed at isolation as well as further characterization of a specific metalloprotein, and therefore involve protein separation, followed by tracing the protein through the presence of the metal. Such an approach was followed by previous investigators in the study of copper and zinc proteins. Richterich[48] used electrophoresis chromatography followed by colorimetric determination of copper in human serum. Parisi and Vallee[49] studied the zinc proteins in serum by DEAE cellulose and gel filtration chromatography, followed by zinc determination using flame atomic spectrophotometry. In addition, on two samples, magnesium, manganese, iron, copper, chromium, and nickel were also measured.

Metalloenzymes and metalloproteins in general, contain only a few metal atoms per

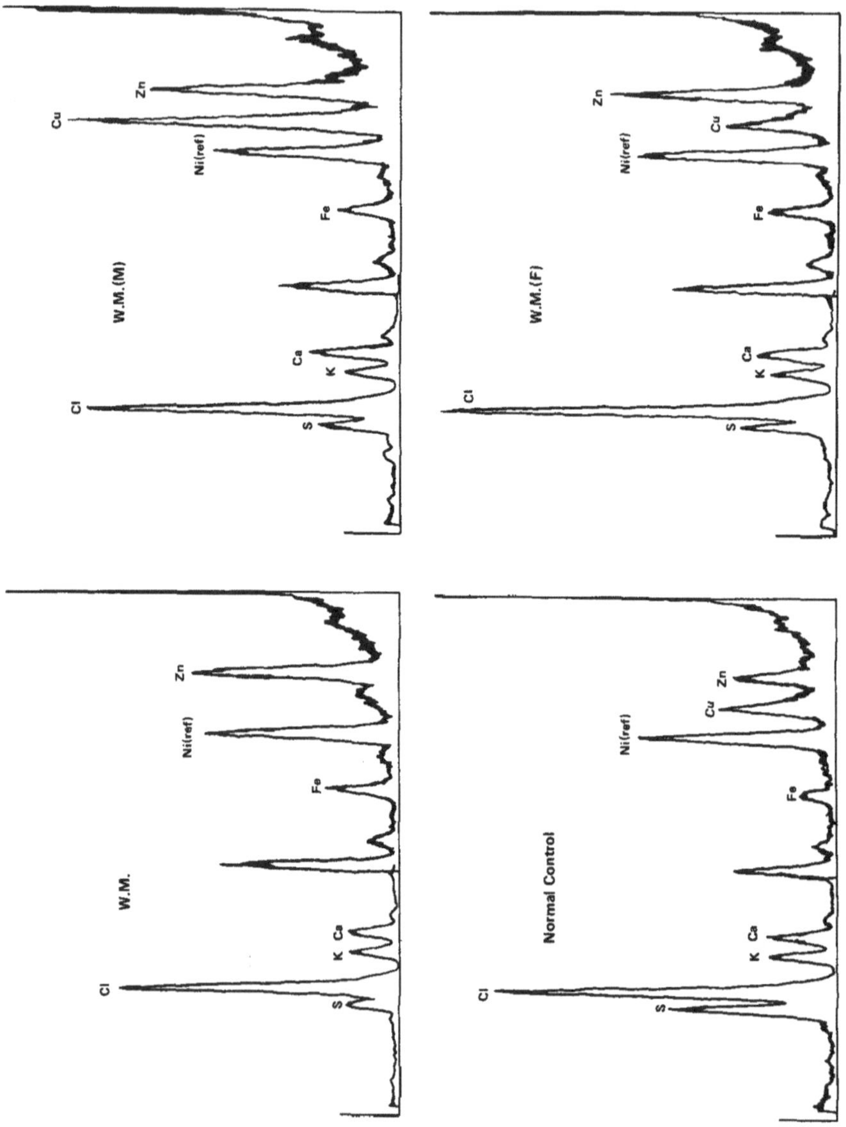

FIGURE 8. Trace elements profile of a patient (WM) with Menke's Kinky Hair Syndrome, his mother (M), his father (F), and a normal control. Note the absence of copper in the boy's serum and the increased amount in his mother's serum.

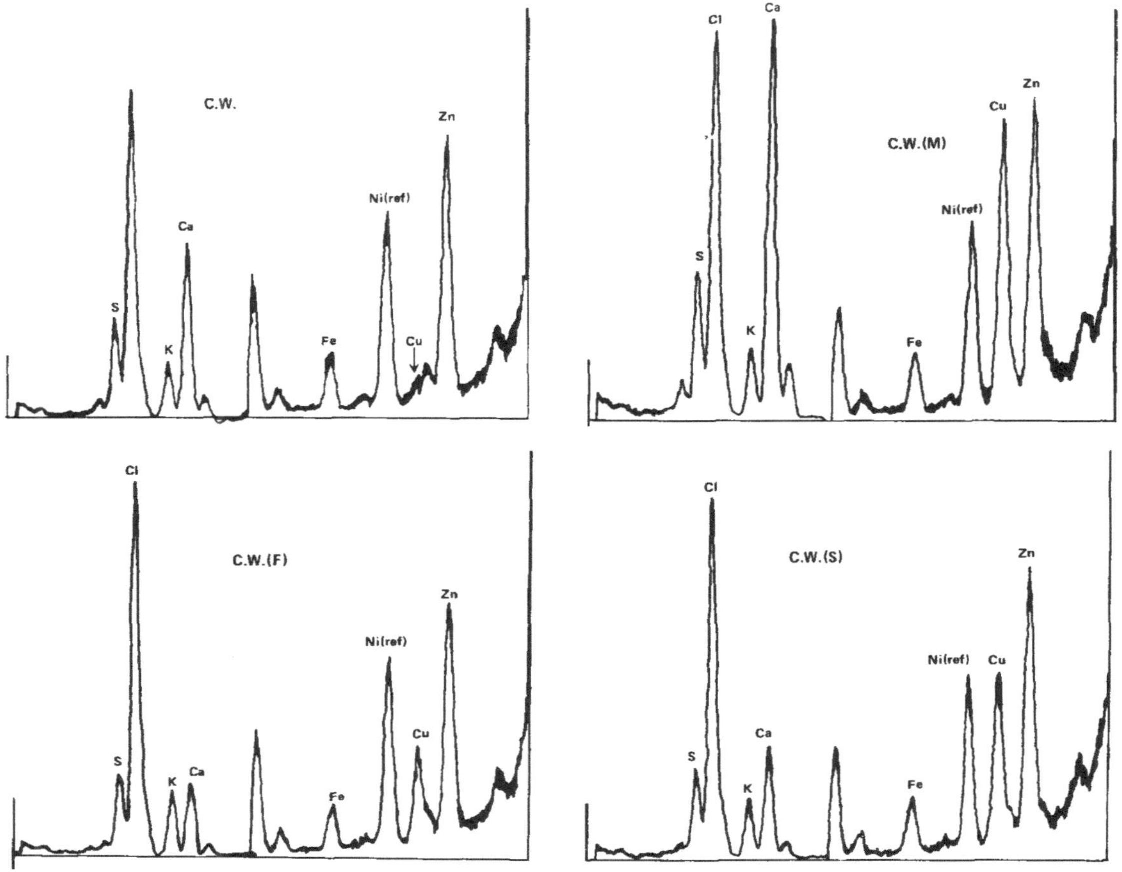

FIGURE 9. Trace elements profile of a patient (CW) with Menke's Kinky Hair Syndrome, his mother (M), father (F), and sister (S). Compare with Figure 8.

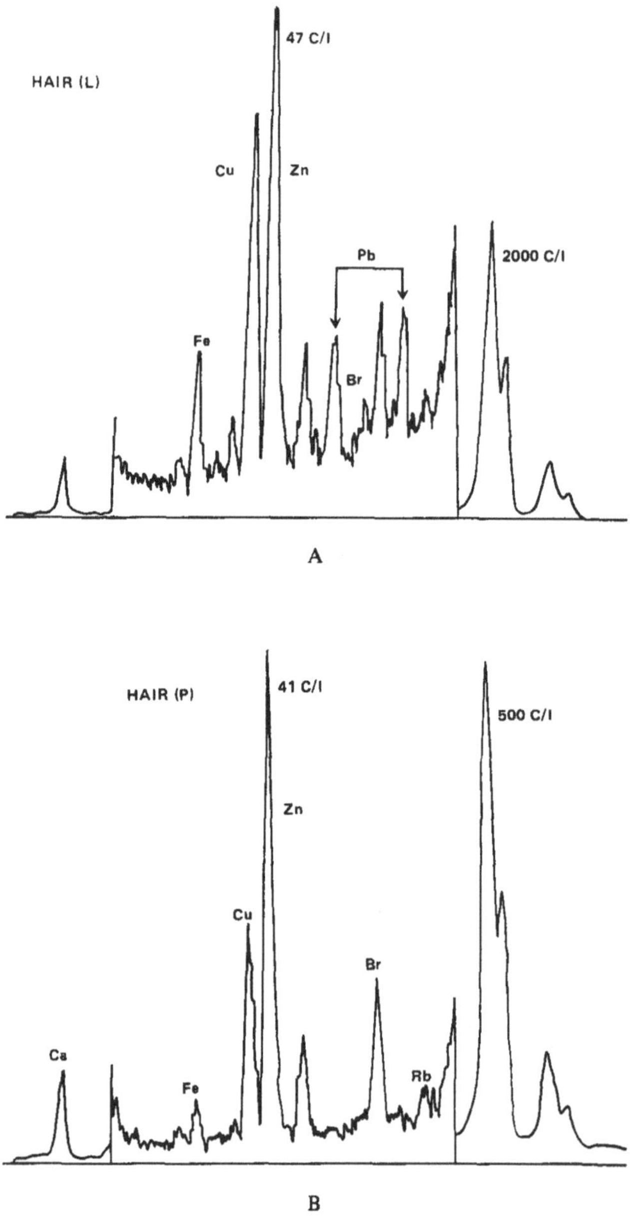

FIGURE 10. Trace element distribution in the hair of a girl (L) and that of her father (P).

molecule. With a molecular weight of about 100,000, the metal concentration is in the order of 100 ppm. The total amount of trace metals present in normal blood serum is, however, very small. For example, about 1.0 μg/mℓ for copper and zinc, 0.1 μg/mℓ selenium, 0.003 μg/mℓ for nickel, and 0.04 μg/mℓ for vanadium. Assuming that all metals are protein-bound, sensitivity in the subnanogram region would be necessary for detection of metalloproteins containing nickel spread over many components. Because of the limited amount of sample available, this type of work would require that the technique be nondestructive or multielemental, with a combination of both having a distinct advantage. The FRIEDA system meets these requirements.

A method of column separation followed by trace element analysis was initiated.

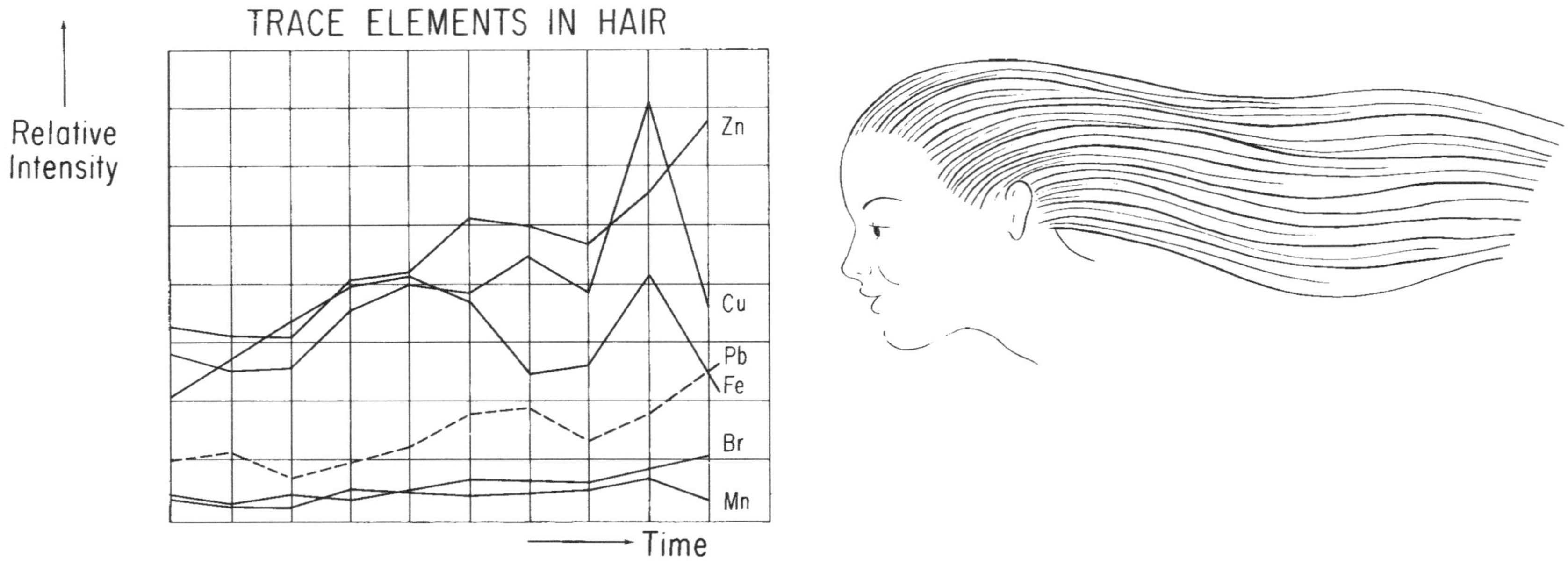

FIGURE 11. The distribution of several elements along the length of the girl's hair.

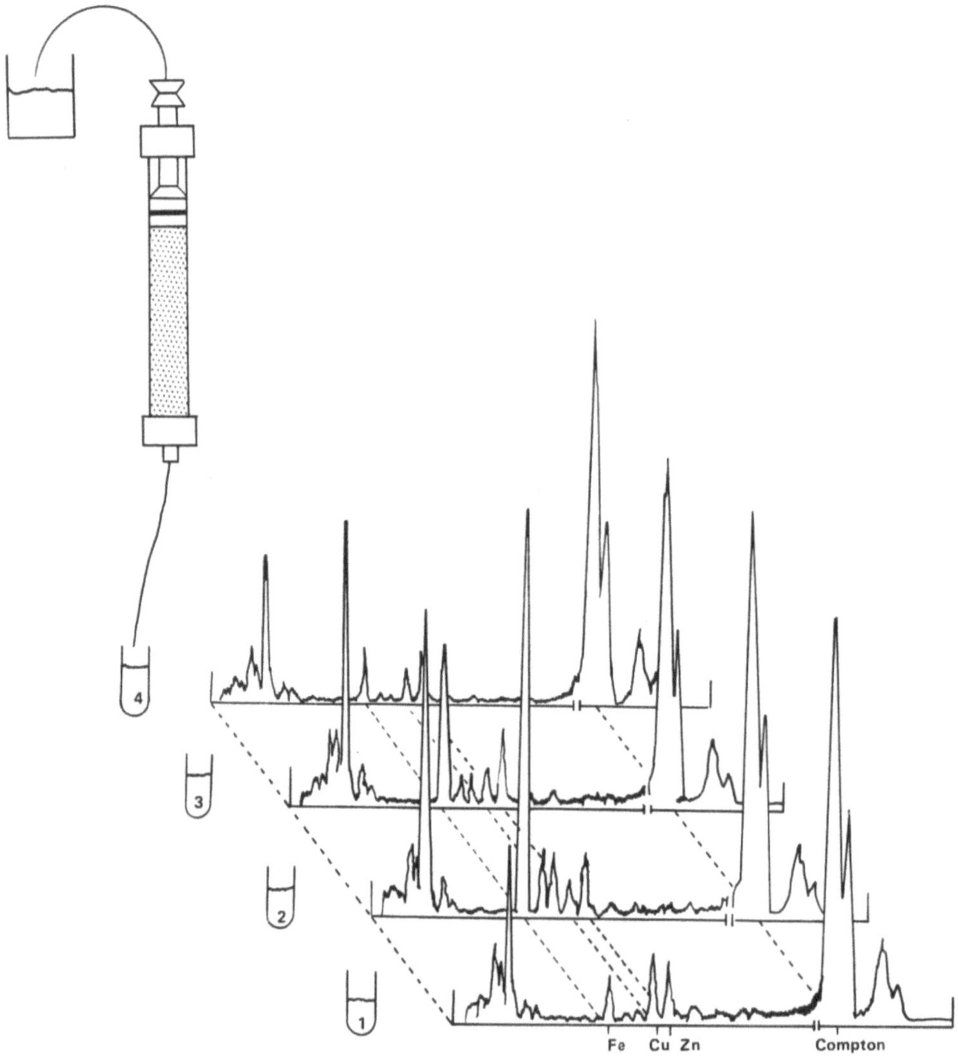

FIGURE 12. Schematic presentation of a two-step metalloprotein study. Serum is first fractionated into its protein components which are subsequently analyzed for their metal content. The resulting spectra may be presented in a three-dimensional metalloprotein profile as depicted.

Two types of columns were used — a Sephadex G200 and a Sephadex DEAE A50. The sample consisted of a 2 mℓ serum from various sources, dialyzed against the appropriate buffer and separated in the column. The fractions were subsequently dialyzed against distilled water and lyophilized. The dried proteins were vacuum sealed in ultrathin polypropylene and analyzed with the FRIEDA system for their trace elements content (Figure 12). The ability of fluorescent techniques for performing this type of analysis in a rapid and reliable manner may prove of use in aiding in the study of the role of metalloprotein in biology.

REFERENCES

1. Bowen, H. J. M., *The Elementary Composition of Mammalian Blood*, United Kingdom Atomic Energy Authority, Res. Group Report # AERE R4196, U. K. Atomic Energy Authority, London, 1963.
2. Mason, B., *Principle of Geochemistry*, John Wiley & Sons, New York, 1952.
3. Dittmer, D. S., Ed., *Blood and Other Body Fluids*, Federation of American Societies for Experimental Biology, Washington, D.C., 21, 1961.
4. Sverdrup, Johnson, and Fleming, *The Oceans*, Prentice Hall, Englewood Cliffs, N.J., 1942.
5. Schroeder, H. A. and Nason, A. P., Trace-element analysis in Clinical Chemistry, *Clin. Chem.*, 17(6), 461, 1971.
6. Sargent, F., Man-environment-problem for public health, *Am. J. Public Health*, 62, 628, 1972.
7. Schroeder, H. A., Metals in the air, *Environment*, 13(8), 18, 1971.
8. Cholak, J., Schafer, L. J., and Yeater, D., The air transport of lead compounds present in automobile exhaust gases, *Am. Ind. Hyg. Assoc. J.*, 29, 562, 1968.
9. Underwood, E. J., *Trace Elements in Human and Animal Nutrition*, Academic Press, 3rd ed., New York, 1971.
10. Bernard, C., Lecons Sur Les Effects Des Substance Toxiques Et Medicamenteuses, Bailliere, Paris, 1857.
11. Warburg, O. and Krebs, H. A., Uber Locker Gebundenes Kupfer and Eisen im Blutserum, *Biochem. Z.*, 190, 143, 1927.
12. Locke, A., Main, E. R., and Rasbash, D. O., The copper and nonhemoglobinous iron content of the blood serum in diseases, *J. Clin. Invest.*, 11, 527, 1932.
13. Sachs, A., Levine, V. E., and Fabian, A. H., Copper and iron in human blood, I. normal men and women, *Arch. Int. Med.*, 55, 227, 1935.
14. Heilmeyer, L. Keiderling, W., and Stiewe, G., *Kupfer und Eisen Als Korpereigene Wirkstoffe Und Thre Bedentung Bein Krankheitsgeschehen*, Jena, Fisher, 1941.
15. Cartwright, G. E., Hugulry, C. M., Ashendrucker, A., Fay, J., and Wintrobe, W. W., Studies on free erythrocytes photoporphyrin, plasma iron and plasma copper in normal and anemic subjects, *Blood*, 3, 501-525, 1945.
16. Pagliardi, E. and Giangrandi, E., Clinical significance of the blood copper in Hodgkin's disease, *Acta Haematol.*, 24, 201, 1960.
17. Jensen, K. B. and Thorling, E. B., Serum copper in Hodgkin's disease, *Scand. J. Haematol.*, 1, 63, 1964.
18. Cherry, N. H., Kalas, J. P., and Zarafonetis, C. J. D.,Study of Plasma Copper Levels in Patients with Acute Leukemia, J. Albert Einstein Med. Center, 9(24), 32, 1961.
19. Hrgovcic, M., Tessmer, C. F., Minckler, T. M., Mosier, B., and Taylor, G. H., Serum copper levels in lymphomas with special reference to Hodgkin's disease, *Cancer*, 21(4), 754, 1968.
20. Hrgovcic, M., Tessmer, C. F., Thomas, F. B., Fuller, L. M., Gamble, J. F., and Shullenberger, C. C., Significance of serum copper levels in adult patients with Hodgkin's disease, *Cancer*, 31(6), 1337, 1973.
21. Tessmer, C. F., Krohn, W., Johnson, D., Thomas, F. B., Hrgovcic, M., and Brown, B., Serum copper in children (6-12 Years): an age correction factor, *Am. J. Clin. Pathol.*, 60, 870, 1973.
22. Cummings, J. N., The effect of B.A.L. in hepatolenticular degeneration, *Brain*, 74, 10, 1951.
23. Craver, L. F.,Hodgkin's Disease, in Tice-Harvey Practice of Medicine, F. Tice, Ed., 6:1017-1064, M. W. Prior, Hagerstown, Md., 1964.
24. Adelstein, S. J. and Vallee, B. L., Copper metabolism in man, *N. Engl. J. Med.*, 265, 892, 1961.
25. Rosner, F. and Gorfien, P. C., Erythrocyte and plasma zinc and magnesium levels in health and disease, *J. Lab. Clin. Med.*, 72(2), 213, 1968.
26. McNeely, M. D., Sunderman, F. W., Jr., Nechay, M. W., and Levine, H., Abnormal concentrations of nickel in serum in cases of myocardial infarction, stroke, burns, hepatic cirrhosis, and uremia, *Clin. Chem.*, 17(11), 1123, 1971.
27. Jaffee, N. and Bishop, Y. M. M., The serum iron level, hematocrit, sedimentation rate, and leukocyte alkaline phosphatase level in pediatric patients with Hodgkin's disease, *Cancer*, 26, 332, 1970.
28. Brendstrup, P., Serum copper, serum iron and total iron binding capacity of serum in acute and chronic infections, *Acta Med. Scand.*, 145, 315, 1953.
29. Elvehjem, C. A., The biological significance of copper and its relation to iron metabolism, *Physiol. Rev.*, 15, 471, 1935.
30. Sinha, S. N. and Gabrieli, E. R., Serum copper and zinc levels in various pathologic conditions, *Am. J. Clin. Pathol.*, 54, 570, 1970.
31. Sunderman, F. W., Jr., Measurements of nickel in biological materials by atomic absorption spectrometry, *Am. J. Clin. Pathol.*, 54, 570, 1970.

32. **Perry, H. M., Tipton, I. H., Schroeder, H. A., Steiner, R. L., and Cook, M. H.,** Variation in the concentration of cadmium in human kidney as a function of age and geographic origin, *J. Chronic Dis.*, 14, 259, 1961.
33. **Tipton, I. H.,** Trace elements in human tissue, *Clin. Nutr.*, 23, 123, 1970.
34. **Tipton, I. H. and Cook, M. J.,** Trace elements in human tissue, Part II. Adult subjects from the United States, *Health Phys.*, 9, 103, 1963.
35. **Tipton, I. H., Schroeder, H. A., Perry, H. M., Jr. and Cook, M. J.,** Trace elements in human tissue, Part III. Subjects from Africa, The Near and Far East and Europe, *Health Phys.*, 11, 403, 1965.
36. **Danielsen, A. and Steinnes, E.,** A study of some selected trace elements in normal and cancerous tissue by neutron activation analysis, *J. Nucl. Med.*, 11, 261, 1970.
37. **Mulay, I. L., Roy, R., Knox, B. E., Suhr, N. H., and Delaney, W. E.,** Trace metal analysis of cancerous and non-cancerous human tissues, *J. Natl. Cancer Inst.*, 47, 1, 1971.
38. **Janes, J. M., McCall, J. T., and Elveback, L. R.,** Trace metals in human osteogenic sarcoma, *Mayo Clin. Proc.*, 47, 476, 1972.
39. **Klevay, L. M.,** Hair as a biopsy material, *Am. J. Clin. Nutr.*, 23, 284, 1970.
40. **Strain, W. H., Steadman, L. T., Lankan, C. A., Jr., Berliner, W. P., and Pories, W. J.,** Analysis of zinc levels in hair for the diagnosis of zinc deficiency in man, *J. Lab. Clin. Med.*, 68(2), 2441, 1966.
41. **Alfrey, A. C., Nunnelley, L. L., Rudolph, H., and Smythe, W. R.,** Medical applications of a small sample X-ray fluorescence system, in *Advances in X-ray Analysis*, Vol. 19, Kendall-Hunt, Dubuque, Ia, 1976.
42. **Bagchi, R. B. K. N. and Ganguly, H. D.,** Lead in Human Tissues, *Ind. J. Med. Res.*, 16(4), 935, 1939.
43. **Hilderbrand, D. C. and White, D. H.,** Trace-element analysis in hair: an evaluation, *Clin. Chem.*, 20, 148, 1974.
44. **Rice, E. W. and Goldstein, N. P.,** Copper content of hair and nails in Wilson's disease, *Metab. Clin. Exp.*, 10, 1085, 1961.
45. **Ong, P. S., Lund, P. K., Litton, C. E., and Mitchell, B. A.,** An energy dispersive system for the analysis of trace elements in human blood serum, 21st Annual Denver X-ray Conf., in *Advances in X-ray Analysis*, Vol. 16, Birks, L. S., Barrett, C. S., Newkirk, J. B., Ruud, C. O., Eds., Plenum Press, New York, 1972, 129.
46. **Ong, P. S. and Cox, H. L., Jr.:** Line focusing X-ray monochromator for the analysis of trace elements in biological specimens, *Med. Phys.*, 3, 74, 1976.
47. **Cox, H. L., Jr. and Ong, P. S.,** Sample mass determination using compton and total scattered excitation radiation for energy dispersive X-ray fluorescent analysis, *Med. Phys.*, 4, 99, 1977.
48. **Richterich, R., Temperli, A., and Aebi, H.,** Die Heterogenitat Des Caeruloplasmins: Isolierung und Charakterisierung Von Zwei Cupro-Proteinen Aus Human-Serum, *Biochim. Biophys.*, 56, 240, 1962.
49. **Parisi, A. F. and Vallee, B. L.,** Isolation of a zinc 2-microglubulin from human serum, *Biochemistry*, 9, 12, 1970.

Section III
Stable Tracers

Chapter 5

TECHNIQUES FOR IN VITRO FLUORESCENT EXCITATION ANALYSIS OF STABLE TRACERS

Leon Kaufman

TABLE OF CONTENTS

I. Introduction ... 91

II. Equipment for In Vitro FEA ... 92
- A. Principles ... 92
  - 1. Limitations Due to Background ... 92
  - 2. Count Rate Limitations ... 93
- B. Detectors ... 94
- C. Electronics ... 95
- D. Sources ... 95
- E. Automated Sample Changer ... 96
- F. Data Processing Electronics ... 96

III. System Performance ... 98
- A. Background ... 98
- B. Quantitative ... 99

IV. Improved Quantitative Techniques ... 103
- A. Lowered Background ... 103
- B. Data Rate Increase ... 104
- C. Compton Suppression ... 105
- D. Polarized FEA ... 105
  - 1. Geometry for Polarized Beam FEA ... 105
  - 2. Equipment ... 105
  - 3. Target Materials ... 106
  - 4. Formulation ... 107
  - 5. Results ... 108
  - 6. Considerations on Polarized Beam FEA ... 109

V. Conclusions ... 113

References ... 113

I. INTRODUCTION

Radio-tracer techniques have gained acceptance in the medical field because of the convenience with which photon-emitting tracers can be identified and quantitated by relatively untrained personnel. On the other hand, the use of radioactive elements presents problems associated with radiation exposure of patients and attending personnel, the need for special handling procedures, the finite life of what are often expensive radio-labeled compounds, and the need for special isolation and disposal procedures when these tracers are used in animals. Fluorescent excitation analysis (FEA) of stable tracers can provide the convenience of radio-tracer technology while avoiding the problems presented above. The main disadvantage with FEA is its relative insensitivity: while nanogram to picogram amounts of radioactive tracers can be readily quantitated, presently it is difficult to do the same with unprepared samples that contain less than a microgram of stable tracer. Even so, sensitivity is not a limiting factor in a number of important applications, and FEA of stable tracers has become the method of choice for performing a variety of clinical and investigative studies, and is now a technology that is diffusing through the medical community both in this country and abroad.

II. EQUIPMENT FOR IN VITRO FEA

The concept behind the equipment developed for medical work is the ability to accurately quantitate tracer concentrations in unprepared biological liquids and tissues. To this purpose, a fully automated system was developed, incorporating design trade-offs intended to maximize its performance in this role.[1] The principles, hardware, and detectors are discussed below.

A. Principles

As discussed in Chapter 1, when an element in a sample is exposed to X- or gamma-rays, the energy of which is higher than the binding energy (BE) of the K-shell electrons in that element, vacancies can occur in the K shells. An L- or M-shell electron will then occupy the K-shell vacancy. In this process, the atom will emit a photon of energy $E\alpha$ or $E\beta$, given by $E\alpha = BE(K) - BE(L)$ and $E\beta = BE(K) - BE(M)$. These photons are called the $K\alpha$ and $K\beta$ characteristic radiations and are unique to each element. Their quantities reflect the amount of a given element present. Similarly, lower energy characteristic X-rays are obtained by ionization of L-shell electrons. Since the excitation probability decreases rapidly as the energy of the exciting photons increases beyond threshold or binding energy ($\sim E^{-3}$), it becomes desirable to use photons close in energy to BE(K) or BE(L) depending on whether K- or L-shell excitation is desired. However, if the energy is too close to the binding energy, Compton scattered photons overlap the characteristic X-rays and the signal-to-background ratio decreases. Compton scattering results in secondary photons of an energy lower than that of the exciting beam, the exact value of that energy depending on the scattering angle. Multiple Compton scattering also produces a diffuse background of lower energy X-rays. Scattered photons have two deleterious effects on the accuracy with which elemental concentrations are measured:

a. They produce background counts in the characteristic X-ray channel of interest, thus degrading the signal-to-noise ratio.
b. They limit the speed with which characteristic X-rays can be detected by occupying the electronics with events that are not of interest.

1. Limitations Due to Background

Referring to Figure 1, in which a schematic drawing of a typical fluorescent spectrum of a tracer-containing sample is shown, we notice the following features: coherent scatter (no energy loss) from the sample; Compton scatter from the sample; the effects of these latter photons undergoing further Compton scatter in the detector; a degraded energy multiple-scatter region (mostly from inside the detector); the characteristic X-rays from the tracer; and background, i.e., nonfluorescence photons in the characteristic X-ray energy window.

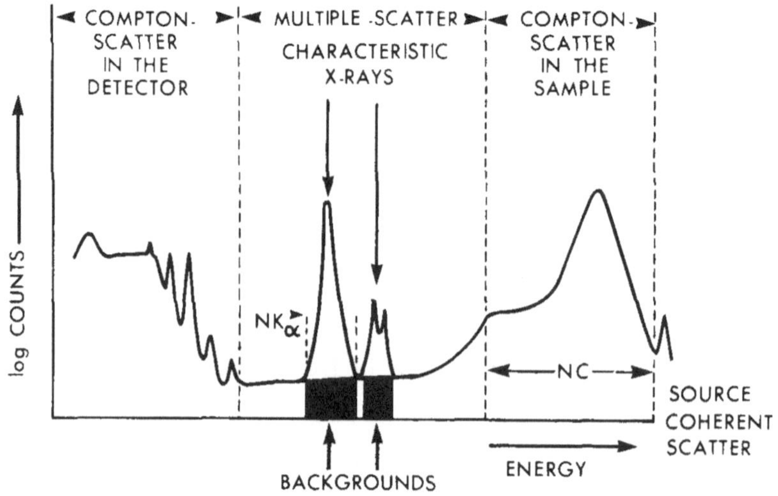

FIGURE 1. Typical FEA spectrum of an iodine-containing sample, excited by Am-241 in a 90-degree geometry.

We have chosen to measure the tracer background in absolute terms, in parts per million (ppm), and the measurement is obtained as follows: defining NK as the counts in the Kα characteristic X-ray channel, NC as the counts in an arbitrary region of the Compton scatter spectrum, B as NK/NC for a sample of water, and W the concentration of tracer in a sample, then

$$W = K\left(\frac{NK}{NC} - B\right) \quad (1)$$

where K is a constant obtained from measuring NK and NC for a sample of known concentration W. From Equation 1, the background in the units of K is KB, and is corrected for dead-time and sample volume variations. KB can be thought of as the tracer concentration necessary to double the counts NK when compared to NK for water alone.

It is usual to state that the statistical validity of counting is determined by the counts in the channels of interest, thus, the percent error would be $\Delta = 100/\sqrt{NK}$. Since NK is linear in time, reproducibility would then increase as $\sqrt{t}$. However, this is only true if $KB \ll W$, that is, if the background is negligible compared to the concentration of interest. If $KB > W$, it is easy to show from equation 1 that Δ is multiplied by a term $\sqrt{(1 + KB/W)}$, that is, the statistics are worsened by a term containing the ratio of background to concentration. We can then write the uncertainty in W (ΔW) due to statistics as

$$\Delta W = \frac{K}{NC}\sqrt{NK} \quad (2)$$

Because the ratio KB to W can reach very large values, reducing background is of paramount importance in accurately quantitating very low level tracer concentrations.

2. Count Rate Limitations

Since the statistics of counting any one sample can be improved by obtaining more counts, it would seem desirable to use stronger sources of excitation, as well as large area, high efficiency detectors to increase the count rate and statistics in a fixed time.

In practice, currently available amplifiers give rise to degraded detector resolution at count rates above 20,000-counts/sec. For low concentrations of tracer, in the best detector systems, no more than 1 out of every 1000 detected events is in the Kα channel. Thus, fluorescence sources cannot be made more intense, since Compton scattered X-rays will eventually set a limit on data rates.

In summary, quantitation of low levels of tracer is limited by two factors: background under the peak of interest produces a need for an increased number of counts, while data rate limitations in the electronics, which are mostly processing scattered photons, do not allow for an increase in the rapidity with which characteristic photons are accumulated.

B. Detectors

Since the tracer signal may represent less than 0.1% of all detected events, high resolution detectors must be used in FEA to separate the characteristic X-rays from other photons. We have investigated the use of lithium-drifted silicon, Si(Li), and high purity germanium, HPGe, for this purpose, and have found that, given a fixed source-sample-detector configuration and esentially equivalent energy resolution, backgrounds among different detectors show variations over a 25-fold range. This variation is not well understood, but it appears to be caused by scattering from materials within the detector assembly of the energetic Compton-scattered photons from the sample, and by intrinsic detector properties. Additional sources of background are crosswalk from the excitation source and multiply-scattered photons in the sample. These latter contributions are not large for a well-designed system, as will be demonstrated later. Since background is a key parameter in the quantitation of low-level tracers, careful consideration of the detector is critical in attaining sensitive systems for medical applications.

The use of HPGe was evaluated for FEA of iodine.Although the absolute detection efficiency of this material is superior to that of silicon, it is as high in the 30 keV region, where the iodine K X-rays are, as at the exciting beam energy (60keV). Thus, dead time introduced by detection of Compton scattered photons decreased the effective detection efficiency of characteristic radiation. In addition, the background was much higher than for silicon in three HPGe detectors obtained from two different manufacturers (Figure 2).

Because of their excellent energy resolution, good signal-to-background ratios, and ready availability, Si(Li) detectors are preferred. Their detection efficiency for K-lines

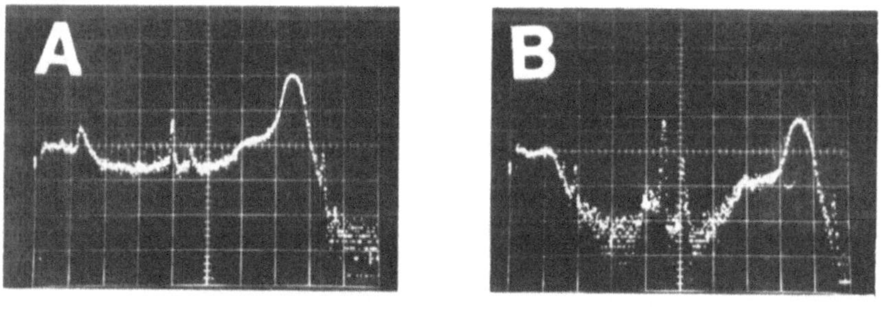

FIGURE 2. FEA spectrum of a sample containing 1 mg/cc of iodine, obtained in 100 sec using the same source-collimator unit. The detectors are (A) an 80 mm², 5 mm-thick HPGe detector with 100% detection efficiency for the iodine characteristic lines, and (B) an 80 mm², 3 mm-thick Si(Li) detector with approximately 60% detection efficiency for the iodine Kα line. In spite of its lower efficiency, the effective count-rate in Si(Li) is higher because of reduced dead time from Compton-scattered photons. Background is also lower in Si(Li) (ordinates are shown in logarithmic scale).

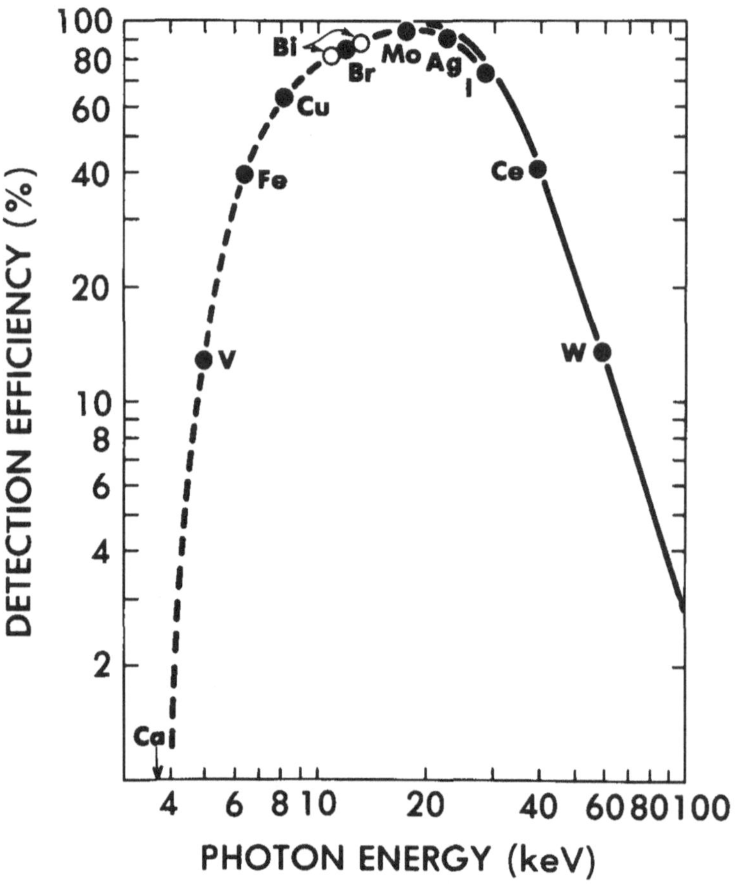

FIGURE 3. Detection efficiency for characteristic lines of various elements for a 5 mm-thick Si(Li) detector. Solid line shows the intrinsic efficiency of the detector, while the dotted line shows the effects of absorption by 1 mm of plastic. Solid dots show the position of the K-lines, while open dots shows the position of the L-lines of bismuth.

of elements reaching up to the rare earths in atomic number is quite adequate (Figure 3). For iodine, the heaviest element known to be a necessary component of the body, the efficiency of a 5 mm-thick Si(Li) detector is over 70% and cannot be much improved through the use of HPGe. Heavy elements can be assayed by using their L-lines for quantitation purposes. At the lower energies, the absorption of even 1 mm of a typical plastic or carbonaceous material makes it difficult to quantitate elements lighter than copper in unprepared samples.

C. Electronics

An important complement to a detector system is its preamplifier and amplifier. In the work described here pulsed-optical-feedback preamplifiers, with reduced gain and short reset times to minimize dead-time, present the most adequate alternative. Shaping times of 2 μsec have been used throughout, and pulse pile-up rejection appears to be a valuable addition to the amplifier.

D. Sources

Since the purpose of our work has been to handle bulk samples without prior preparation such as ashing or slicing, disposable 2-mℓ plastic vials were chosen as sample

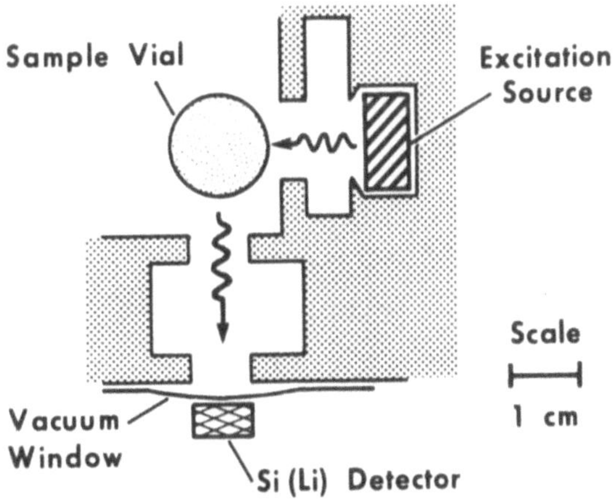

FIGURE 4. Arrangement for FEA of unprepared samples in a 90-deg geometry.

holders. These typically contain between 1 and 1.75 mℓ of sample. To minimize Compton scattering, a 90-degree geometry is chosen between the excitation sources and the detector (Figure 4). Both source and detector are collimated to 1 cm-diameter, with a substantial penumbera.

Although X-ray tubes in various configurations (Chapter III) can yield very large photon fluxes, radioactive sources are available which can saturate detector data rates when used with samples of the size mentioned above. Because these sources are compact, highly reliable, and cheap when compared to X-ray units, our work has concentrated on their use. Am-241 (60 keV, 500-year half-life) and Cd-109 (22 keV, 0.75-year half-life) are excellent sources to cover the detection range made available by Si(Li) detectors. We are presently using a 1.2 Ci Am-241 source. It is filtered with 0.0025 cm of hafnium, to decrease the emission of lower energy and multiply-scattered photons while providing minimum absorption for the 60 keV gamma-ray used for excitation. By varying the source's position within its collimator, 60 mCi of reactor-produced Cd-109 yield a usable lifetime of 3 or 4 years.

For quantitation of heavy elements through detection of their Kα lines, HPGe detector with Gd-153 or Co-57 excitation sources appear to be the best choice.

To minimize scatter from the collimator walls, they are configured as black holes to present a minimum scattering cross-section (Figure 5). To further reduce collimator scatter, the Am-241 collimator is made of Mallory 1000, a tungsten alloy, since L-line fluorescence induced in this material by the source has an energy lower than that of the K-lines of tracers that are fluoresced with Am-241. For similar reasons, cadmium collimators are used with Cd-109. In both cases, the Compton-scattering cross-section in the collimator walls is low compared to the dominant L-edge absorption.

E. Automated Sample Changer

A sample changer was designed to present the disposable vials to the fluorescent beam in such a manner that the 90-deg geometry is maintained (Figures 6 and 7). The capacity of the changer is 48 samples. To obtain a homogeneous distribution of tracer in samples such as heparinized whole blood (which tends to settle), the operator can select an option where samples are mixed just prior to assay.[2]

F. Data Processng Electronics

Data is accumulated in a 2000-channel programmable pulse-height analyzer (PHA) (Figure 6). This unit incorporates a minicomputer behind the front panel. A program

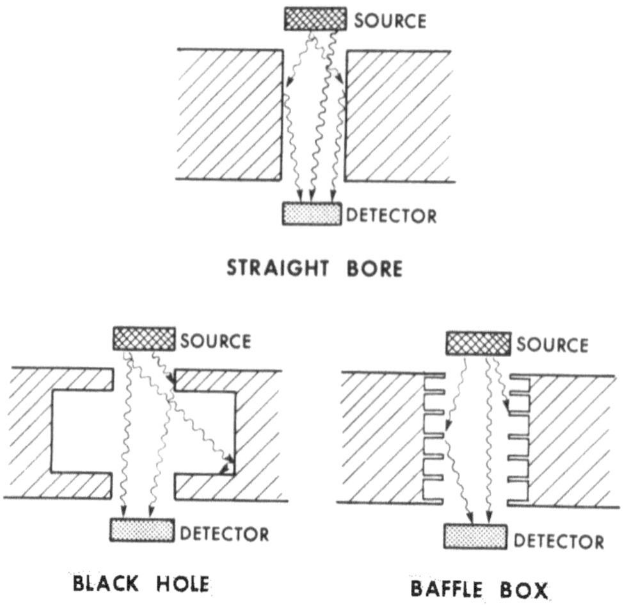

FIGURE 5. When compared to a straight bore, black hole or baffle-box collimator configurations decrease the amount of scattered and fluoresced radiation reaching the detector.

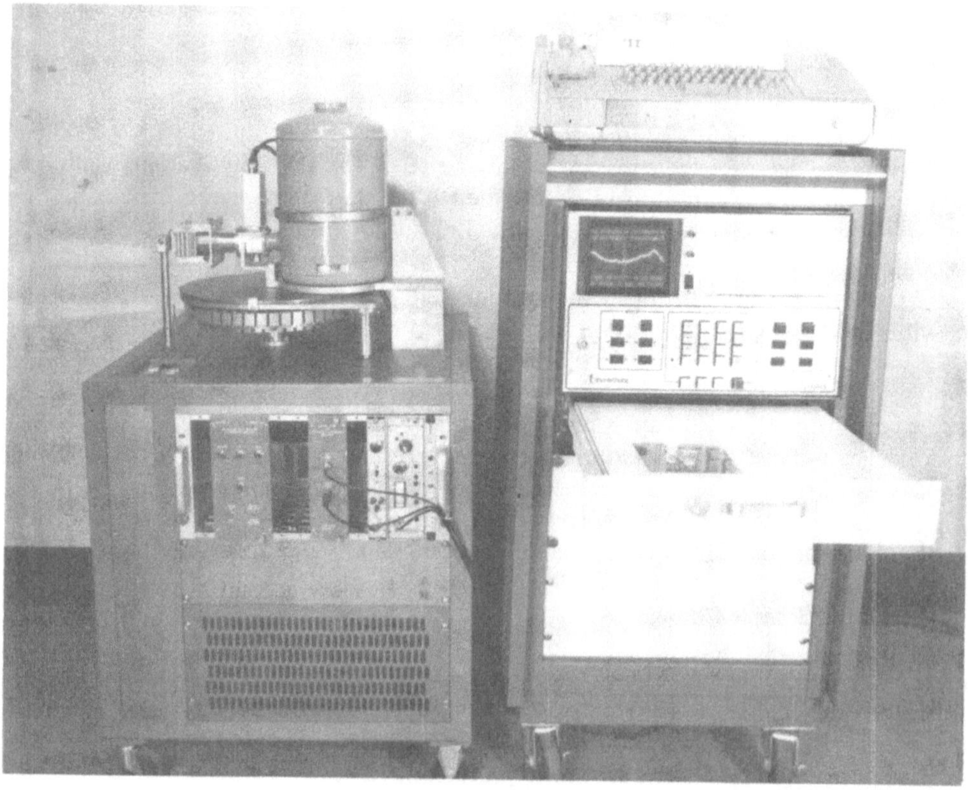

FIGURE 6. Automated FEA System. At left is the sample changer, with detector on top. The arm is in the up position. At right is the programmable PHA. Sample vials are shown on the PHA shelf.

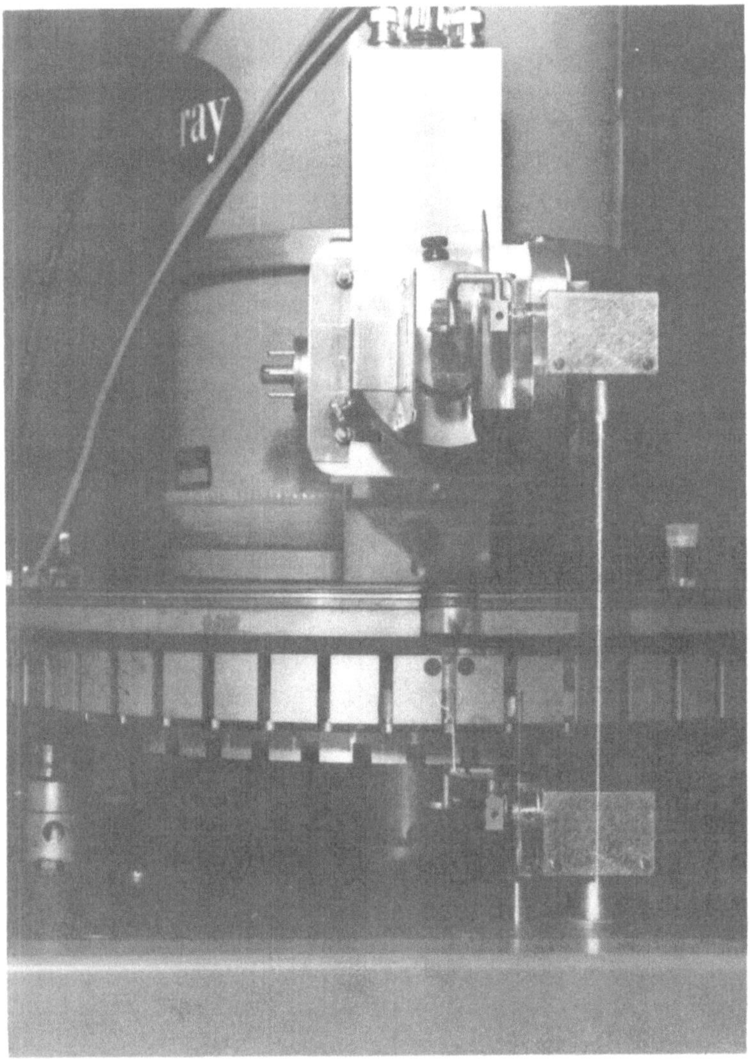

FIGURE 7. Detail of the sample-positioning mechanism. Double exposure shows pick-up arm in the down and up position. Excitation beam reaches the sample from the right, and the detector is behind the collimator hole seen through the sample vial.

tape sets up automatic operation with variable parameters controlled by data entered through a teletype. Accumulation modes of constant time or constant NK can be selected, but because of the wide dynamic range of the system a preferred mode sets minimum and maximum counting time, with accumulation otherwise completed when a preset integral is reached. Printout includes NK and NC, counting time, the concentration W and the statistical error ΔW, where terms are defined as above in Equations 1 and 2, or Equation 3 below. Physiologic parameters of interest, e.g., glomerular filtration rate, red cell volume, extracellular fluid volume, calculated as described in Chapter 6, can also be obtained by entering the appropriate ancillary data.

III. SYSTEM PERFORMANCE

A. Background

As previously discussed, background under the fluorescent peak-of-interest can be

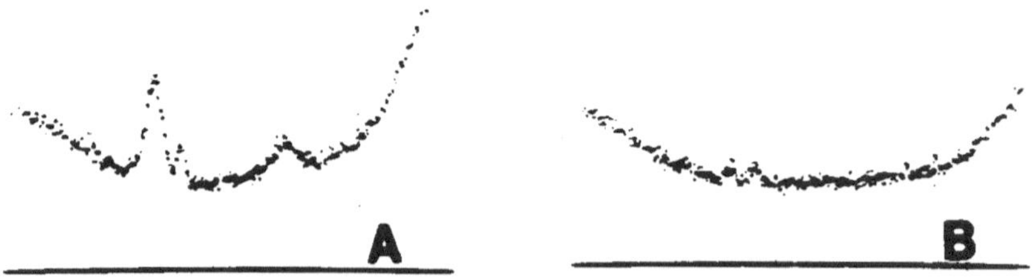

FIGURE 8. FEA spectra of a water sample, obtained with an Am-241 source, normalized to equal peak counts in the Compton peak. The cursors indicate the region between silver and cerium K-lines. (A) The normal spectrum shows peaks from the scatter of low-energy Am-241 lines. (B) The beam reaching the detector has been filtered with 0.076 mm of tantalum, which strongly decreases low-energy photons, as evidenced by the suppression of the Am-241 lines and multiple-scatter to the right of the upper cursor. The overall background decreases by just 12%, indicating that the major contribution is from higher-energy photons degraded in the detector assembly. Cross-talk from the source contributes less than 2% of the counts in this region.

divided into internal and external components. The latter arise from crosstalk from the source (usually low-abundance high energy gamma-rays that are inadequately shielded), from multiple-scatter in the sample, and internally from the detector assembly and detector. These contributions can be evaluated as follows: the count-rate with the source shutter closed yields crosstalk. The total contribution due to the external component of background when Am-241 is used is obtained by filtering the input to the detector with an element such as Hf or Ta, which absorbs the low energy X-rays that would reach the detector more readily than the main Compton peak. Silver can be used for the same purpose with Cd-109. Results in the silver to cesium fluorescence region using Am-241 excitation are shown in Figures 8A and 8B, and were otained with an 80 mm^2, 5 mm-thick, KeVex detector with a pulsed-optical-feedback preamplifier and pulse-pile-up rejection. As can be seen, the major source of background is internal, from the detector and its assembly, and varies little over the energy range of interest.

Since multiple scatter in the sample contributes a significant fraction of background for elements with fluorescent lines that lie nearer to the Compton peak, and the amount of scatter is dependent on sample volume, a variant of equation 1 is used for quantitation, namely

$$W = K [NK - B(W_1 + W_2)] / NC \quad (3)$$

where W, NK, and K are defined as in Equation 1, W_1, and W_2 are the counts in two energy windows just below and just above the peak of interest, and B is $NK/(W_1 + W_2)$ for water. While Equation 3 provides accurate quantitation (Figure 9), the statistics are worsened. Thus, Equation 3 is used for elements such as cerium and lanthanum, where systematic errors due to variable volume would be too large, and not usually for elements such as iodine, cesium and barium, where Equation 1 is adequate (Figure 10).

B. Quantitation

It is customary to define performance in terms of sensitivity, the tracer concentration that yields a net signal equal to three times the square root of the counts for a tracer-free sample, both accumulated in the characteristic X-ray channel in a counting time of 1000 sec. Clearly, sensitivity is a measure of the confidence of detectability of a tracer. In terms of the medical applications discussed here, such a measure has little significance, since maximizing sensitivity does not necessarily maximize the accuracy

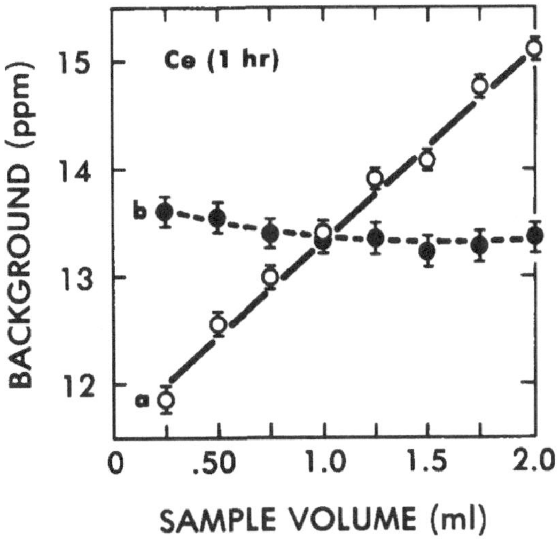

FIGURE 9. Multiple scatter contributes to cerium background, since it varies as a function of sample volume when computed using equation 1(a). For samples with more than 0.50 m*l*, computation using equation 3(b), reduces the variability to within 1 standard deviation for 1 hr measurements.

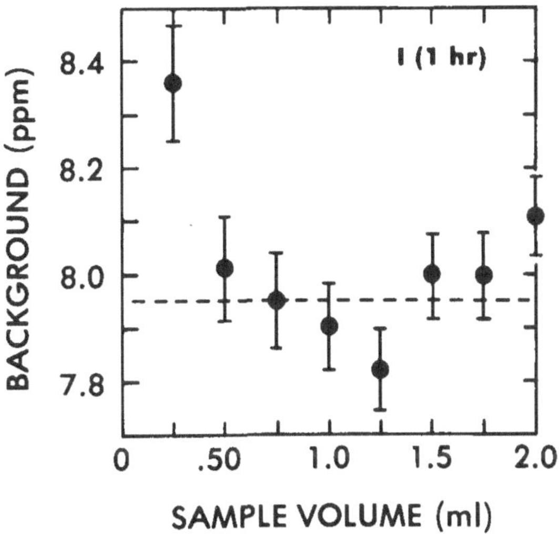

FIGURE 10. Equation 1 is satisfactory for correction of iodine background in samples that have between 0.5 and 1.75 m*l* in volume, since the systematic deviations remain with 1 standard deviation for 1 hr measurements.

of the results obtained from Equations 1 or 3. For instance, Figure 11 shows that the maximum sensitivity occurs at a NK window with a width of 1σ (1σ = full width half maximum, FWHM). On the other hand, maximum accuracy of quantiation is dependent on the tracer concentration range of interest. Figure 12 shows optimum NK window as a function of W/KB, the ratio of tracer concentration to absolute background, and indicates the need for a variable window, which would be easily programmed for. In practice, another parameter must be taken into account: given the drift specifications

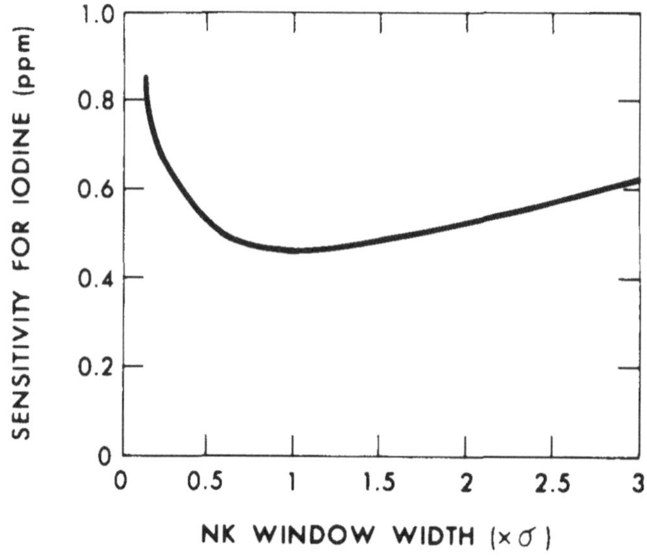

FIGURE 11. Maximum FEA sensitivity for iodine is obtained for an NK window width of 1σ(= FWHM).

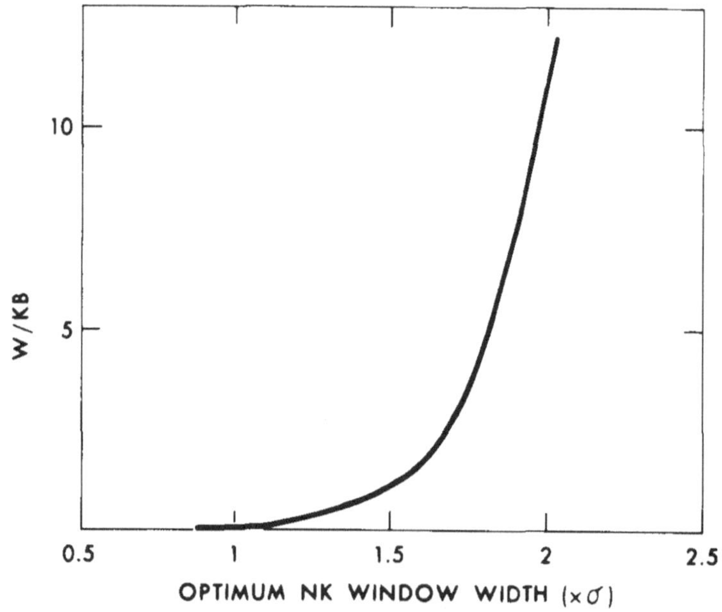

FIGURE 12. The optimum window width for quantitation should be set after obtaining an estimate of concentration from a constant window, since it depends on the ratio of concentration to background (1σ = FWHM).

of present analog to digital converters (ADC), random drifts of at least one channel per 2000 channels of memory cannot be avoided. In our system this represents a drift of 0.083σ. Figure 13 shows the error introduced by shifts of one and two channels. As can be seen, a minimum NK window width of 1.5σ should be used to remain reliably within a 2% margin of error.

In a complex peak such as the Kα line of iodine, which contains a Kα1 line at 28.610 keV and a Kα2 line at 28.315 keV, the NK window should be further broadened to compensate for the separation between the two lines.

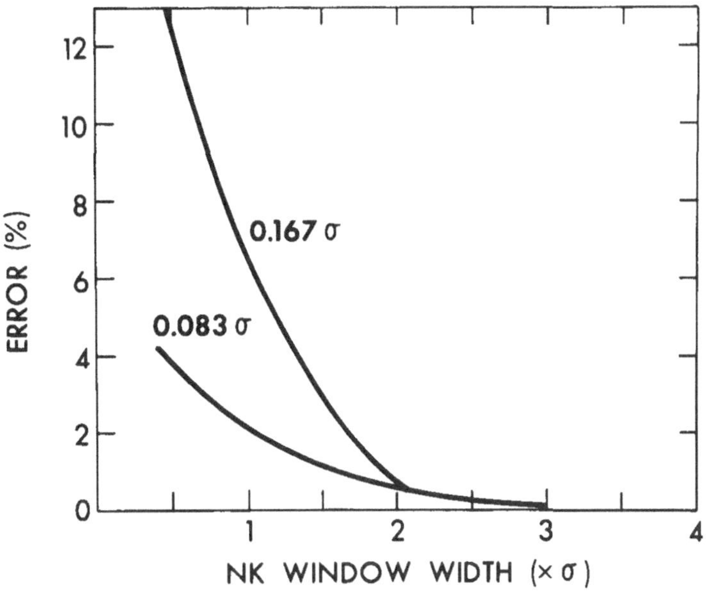

FIGURE 13. Minimum window widths for reliable determination of concentration are limited by random PHA drift. The systematic error introduced in iodine quantitation by a gain or baseline drift of 1 (0.083σ) and 2 (0.167σ) channels per 2800 are shown as a function of NK window width (1σ = FWHM).

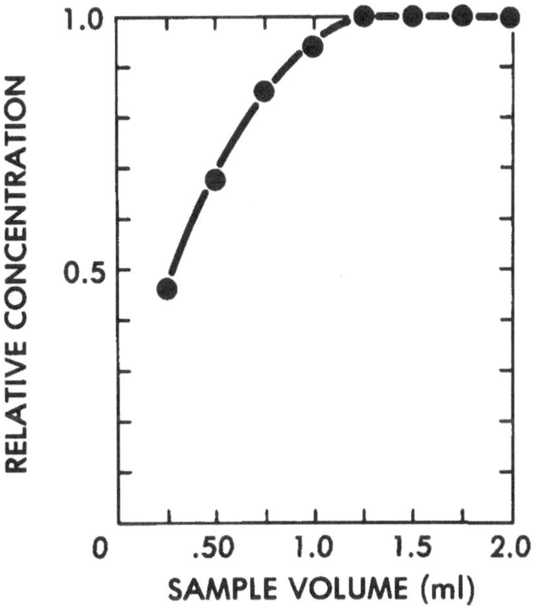

FIGURE 14. To maximize sample irradiation, the exciting beam has a large penumbra. This introduces significant quantitation errors for samples with less than 1 mℓ of volume.

Because the ratio of NK to NC is used to obtain W (Equations 1 and 3), results are independent of dead time effects. The effects of sample volume on background have already been analyzed (Figures 9 and 10), and the effects on concentration determination (that is, on K) are independent of volume once a minimum sample level has been achieved (Figure 14). The response of the system is linear for determining the concen-

TABLE 1

System Performance Parameters

Excitation source	Tracer	Net count-rate (counts/sec/ppm)	Background (ppm)	Sensitivity (ppm)
Cd-109, 38 mCi	Bromine	0.21	13.3	0.76
	Rubidium	0.23	14.5	0.75
Am-241, 1.2 Ci	Silver	0.38	8.07	0.44
	Iodine	0.38	7.95	0.43
	Cesium	0.38	11.75	0.53
	Barium	0.39	11.85	0.52
	Lanthanum	0.36	11.50	0.54
	Cerium	0.35	13.70	0.60

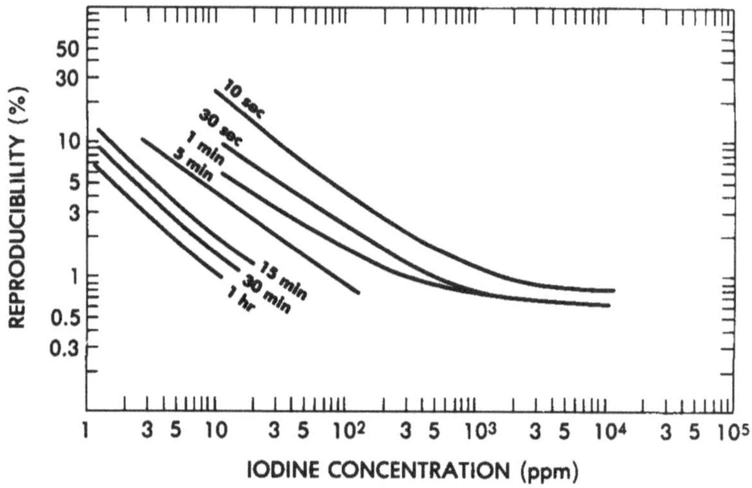

FIGURE 15. The reproducibility with which iodine can be quantitated in the automated FEA system is shown as a function of tracer concentration for various counting times.

trations of elements such as Br, I, and Cs over a broad concentration range, from less than 1 ppm to over 2 to 4% concentration. The upper range is limited by self absorption and electronic degradation in performance at very high counting rates, and the lower limit is determined by counting statistics. This large dynamic range is of importance in medical work, where low concentrations biologic samples have to be compared to higher concentration injectates.

Table 1 summarizes some relevant parameters for our automated FEA system. The accuracy of concentration determinations for some of these elements are shown in Figures 15 to 17.

IV. IMPROVED QUANTITATION TECHNIQUES

A. Lowered Background

Improving the low-tracer quantitation capabilities of FEA is important for extending the applicability of the technique to new areas as described in this and the following chapters.

Based on the data shown in Figure 8, reducing detector-generated background appears to provide the potential for an improvement of a factor of five, clearly significant

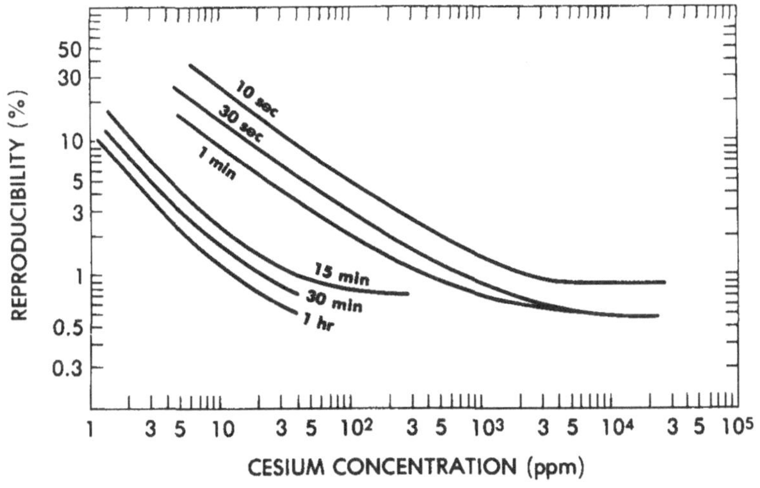

FIGURE 16. Reproducibility as a function of cesium concentration.

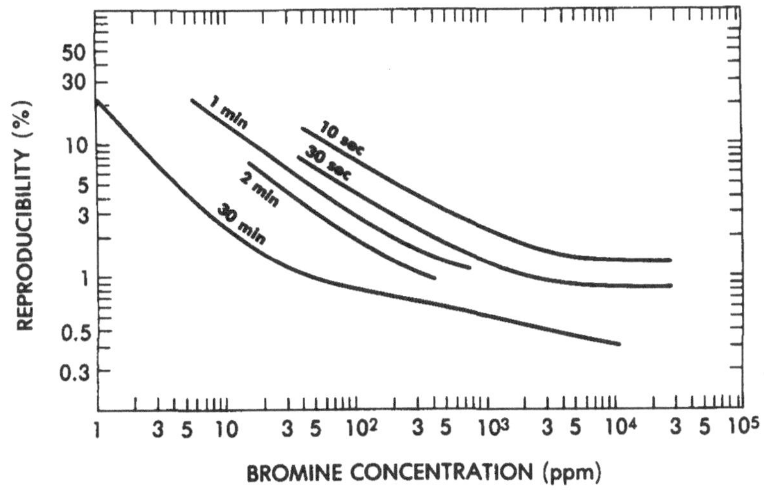

FIGURE 17. Reproducibility of quantitation of bromine using a Cd-109 source which had decayed to an activity of 38 mCi.

for samples with a concentration of under 1 ppm. For instance, the substantial quantitation time of sub-ppm iodine concentrations would be shortened linearly with the decrease in background. This will be partially offset, though, by the need to use Equation 3 for quantitation purposes, since as internal detector background decreases the relative contribution from the multiple-scatter component increases.

B. Data Rate Increase

Improvements in the rate at which data is collected will continue at a slower pace. Amplifiers with increased throughput continue to be developed, but effective improvements are marginal. Table 2 compares the characteristics of a Tennelec 205A BLR to a KeVex 4510P amplifier, both operating with 2 μsec pulse-shaping times, and under identical conditions. While the throughput of the 205A is higher, the lack of a pile-up rejector generates a higher background, which tends to offset the gain in data rate. Another limiting factor is the ADC in the PHA. For a 2000-channel display and a 100

TABLE 2

Comparison of Amplifiers

For Am-241 FEA of Iodine, 80 mm^2 × 5 mm Si(Li)

Amplifier	NK Count-rate (For 1.5 m*l* of 12,700 ppml)	Background (ppml)	Energy resolution (FWHM at 30 keV)
Tennelec 205A BLR Mod.	3286 cts/sec	13.0	238 eV
KeVex 4510 P W/pile-up rejector	2857 cts/sec	12.3	225 eV

MHz ADC, Compton peak photons will require 15 to 20 μsec to be processed. Thus, for an event rate of 20,000 counts/sec, approximately 40% will be lost by the PHA. Faster ADCs, with 200 to 500 MHz clocks are now coming into the marketplace and will improve data accumulation capabilities accordingly.

C. Compton Suppression

Large gains in background and throughput capabilities are possible if Compton scattering in the sample could be eliminated. Conceptually this could yield improvements in speed or minimum quantitation levels that would be as large as 10^4.

One clear way to improve the fluorescent yield to Compton ratio is to lower the energy of the exciting beam. For iodine, a 40 keV beam will be 3 times as effective as a 60 keV beam in eliciting fluorescent X-rays, while the increase in Compton scatter from the matrix will be only of the order of 17%, thus, the fluorescent yield to Compton will be improved by a factor of 2.6. Unfortunately, the overlap of the Compton peak with the fluorescent X-rays will more than negate this gain, since measurements indicate that background increases by a factor of 3 for iodine, and worse for elements above this tracer.

As discussed in Chapters 1 and 3, Compton scatter observed by the detector can be reduced in relation to the fluorescent X-rays by utilizing polarized X-ray beams for excitation, and by locating the detector along a direction of minimum Compton scatter.

D. Polarized FEA

1. Geometry For Polarized Beam FEA

Figure 18 shows the geometry that realizes polarized beam FEA. Unpolarized photons from the X-ray source are collimated along with the z axis, striking a polarizing target. Scattered radiation from the target is collimated along the x-axis and reaches the sample. The scattered radiation thus selected is polarized orthogonally to the xz scattering plane, and cannot incoherently scatter at 90° into its own plane of polarization (the xy plane). Thus, a detector placed along the y-axis will see little or no incoherently scattered radiation from the sample. In practice, the finite solid angle of the collimators and some multiple scattering conspire to produce less than perfect polarization.

2. Equipment

The primary excitation source used[3] is a Philips MG160 X-ray tube and generator

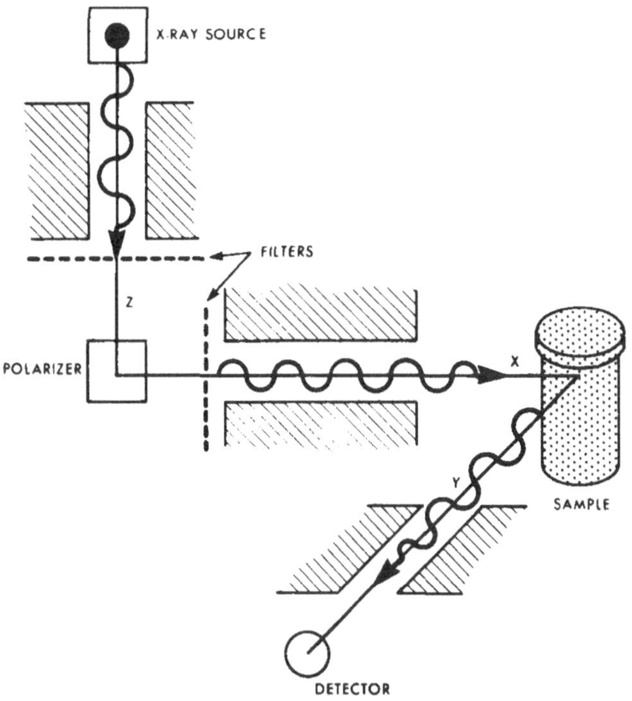

FIGURE 18. Geometric arrangement for minimizing Compton scattering into the detector. The primary beam is unpolarized (z-axis). The excitation beam (x-axis) is polarized in the x-y plane.

capable of continuous operation at 160 kV (constant potential) and 19mA. To avoid the excitation of interfering characteristic lines in the collimators, they are fabricated from electrolytic copper. Three cylindrical boron carbide targets are used, each of 1.2 cm diam and 1 cm length. The axis of the cylinders are aligned towards the sample. Various filters could be introduced in the path of the polarized beam, just after it exits the polarizing targets. The disposable 2 mℓ sample vials are those used in the automated system, and contained approximately 1.5 mℓ of fluid. The spectrometer consists of a 80 mm^2, 5 mm deep low-background KeVex Si(Li) detector with pulsed-optical feedback preamplifier, a Tennelec TC 205A BLR amplifier modified by the addition of a pulsed-optical reset card and operated with a 1 μsec shaping time, and an Ino-Tech, Inc., Ultima II 2000-channel programmable pulse-height-analyzer with 100 MHz ADC.

3. Target Materials

A simple model can be used to evaluate the scattering efficiencies of different polarizing materials.[4] Assume that the incoming beam to, and the outgoing beam from, the target materials have absorption paths of 0.5 cm, and that scattering occurs along a path length of 1.0 cm. The scattering efficiency is then given by the product of the incoming beam attenuation, the outgoing beam attenuation, and the scattering of the incoming beam, i.e.,

$$N = \exp(-\mu_{ti}\rho \times 0.5) \times \exp(-\mu_{to}\rho \times 0.5) \times \{1 - \exp[-(\mu_{ti} - \mu_{pi})\rho \times 1.0]\} \qquad (4)$$

where μ_{ti} and μ_{to} are the total mass absorption coefficients at the incoming and outgoing beam energies, μ_{pi} is the photoelectric mass absorption coefficient at the incoming beam energy, and ϱ is the target material density. For the purpose of obtaining a max-

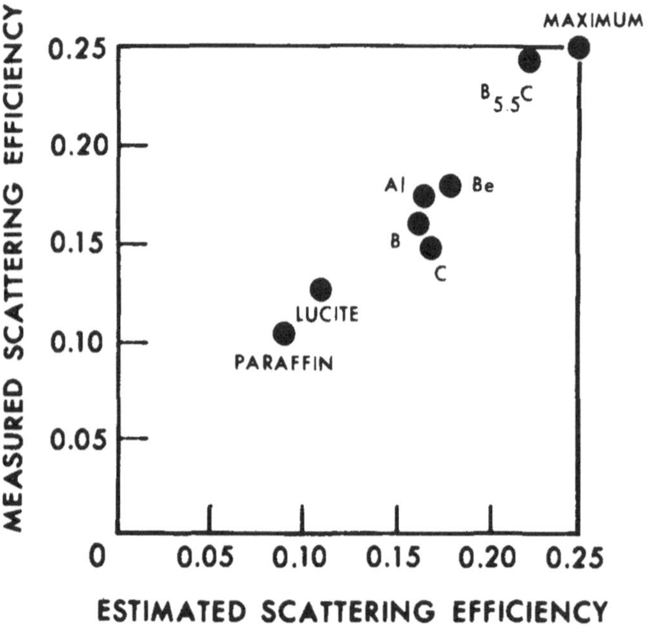

FIGURE 19. Measured and calculated scattering efficiencies of various polarizing targets at 60 keV.

imum value for N, assume that $\mu_{ti} = \mu_{to} = \mu >> \mu_{pi}$, which is a reasonable approximation for low Z materials at 60 keV. Equation 4 becomes

$$N = \exp(-\mu\rho \times 1.0) \times [1 - \exp(-\mu\rho \times 1.0)] \qquad (5)$$

Setting $dN/d(\mu\varrho) = 0$ we obtain

$$\exp(-\mu\rho) = 1/2 \text{ and } N_{max} = 0.25$$

Data on scattering efficiencies were obtained for a number of target materials using an Am-241 source, and are compared to the predicted values as shown in Figure 19. The best material tested was boron carbide ($B_{5.5}C$), which is close to optimum for the target geometry used. Thus, the simple model represented by Equation 4 was found to be adequate for the purpose of predicting the scattering efficiency of various target materials.

4. Formulation

From Equations 1 and 2, the uncertainty with which a tracer can be quantitated can be expressed as

$$\frac{\Delta W}{W} = \left(\frac{1 + KB/W}{NK - B \times NC}\right)^{1/2} \qquad (6)$$

Since $NK = CR \times (W + KB) \times t$, where CR is the count-rate in cts/sec/ppm, the term $(NK - B \times NC)$ is simply $CR \times W \times t$ and

$$\frac{\Delta W}{W} = \left(\frac{1 + KB/W}{CR \times W \times t}\right)^{1/2}$$

Since our interest in the effects of polarized excitation is in applying this mode to the low tracer quantitation regime, i.e., KB>>W, we can write

$$\frac{\Delta W}{W} \simeq \frac{1}{W}\left(\frac{KB}{CR \times t}\right)^{1/2} \tag{7}$$

Thus, it can be seen that for a certain concentration level and constant quantitation accuracy, the counting time is proportional to KB/CR. For low tracer levels, the improvement factor in utilization of the FEA system will be given by the ratio of the time to perform the assay using the source-excited system to the time necessary for obtaining the same quantitation accuracy using the polarized mode. As such, the improvement factor is an index of the increased speed of the system. Note that optimizing this parameter is not necessarily compatible with optimizing sensitivity as defined previously. While the improvement in speed is linearly dependent on KB/CR, the improvement in sensitivity is dependent on $\sqrt{KB/CR}$, which underemphasizes the importance of the background.

5. Results

In the set up for polarized beam FEA there are five parameters that can be varied independently: collimation, tube kilovoltage, filter thickness and material, and distribution of filters between the primary and polarized beams. The first factor was determined by reducing all collimator lengths as much as possible within the limitation imposed by the dimensions of the X-ray tube, sample vial, and detector assembly. In regards to the last factor, we found that filtering the polarized beam only was consistently more effective than distributing the filtration between it and the primary beam.

Filter material was chosen so as not to produce interfering fluorescent X-rays, while trying to maintain as high an atomic number as possible. For the excitation of tracers with atomic numbers above 45, zirconium is a convenient filter material. To reduce its characteristic X-ray contribution to total count-rate, zirconium filters were followed always by a 0.15 mm copper filter.

With collimation and filter material fixed, only two parameters need be varied: filter thickness and tube kilovoltage, studying their effects on KB and CR with the goal of obtaining as large an improvement factor as possible. Because of its ubiquity as a tracer and its biologic importance, most of the work was centered around iodine as the tracer.

The detector-amplifier combination in use had a background of 11.5 ppm I for Am-241 excitation, but the improvement factor was computed with respect to the detector in the automated FEA system, where the background was 8.1 ppm I. NKα window widths were set to 10% of peak value above background, following the criteria described previously.

Figures 20 and 21 show the effects of tube voltage and filter thickness on background and improvement factor of iodine, respectively. It can be seen that background is not strongly dependent on tube voltage, and that there is little change above 0.45 mm Zr filtration. The improvement factor, on the other hand, has a strong dependence on tube voltage, and, while the background changes negligibly between 0.48 and 0.58 mm Zr filtration, the improvement factor was more strongly affected.

Background for 150 KV tube voltage was studied as a function of filter thickness for the elements between tellurium and cerium. As can be seen (Figure 22), while KB does not change much for Te and I, it is strongly dependent on filtration at the higher atomic numbers. Thus, while filtration with more than 0.45 mm Zr is not helpful for iodine, it offers advantages for the higher atomic number elements.

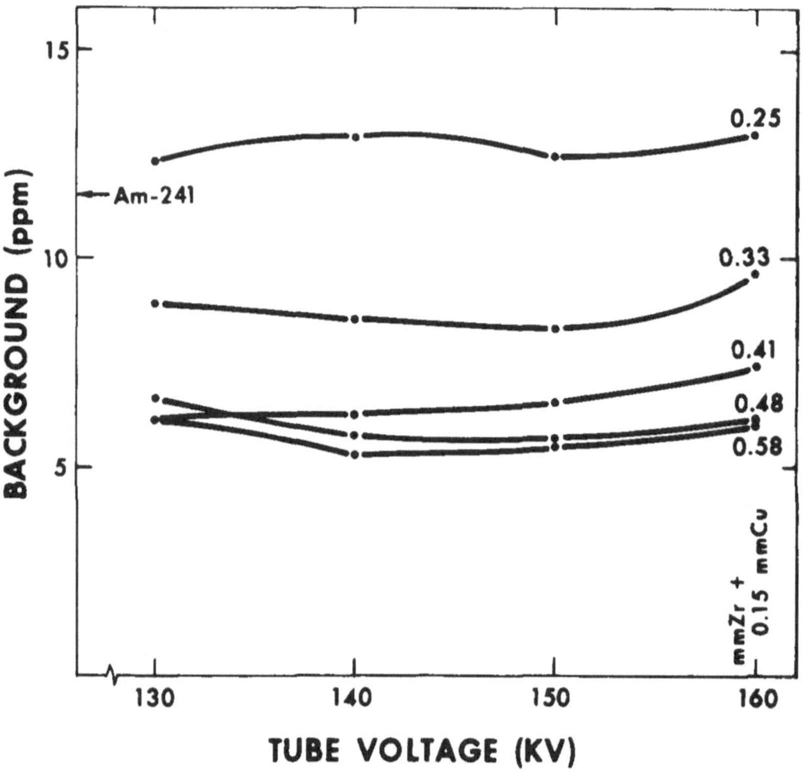

FIGURE 20. Dependence of iodine background on tube voltage for various filter thicknesses. The arrow indicates the background measured for Am-241 FEA in this detector.

The effect of filtration on the improvement factor for various elements is shown in Figure 23. Operation at 150 KV with 0.48 mm Zr plus 0.15 mm of Cu filtration seems optimum for the tracer range shown here, where improvement factors in the range of 4 to 11 are possible. The same system, using copper filtration and operated at 55-65 KV, can realize improvement factors of over 20 for bromine, when compared to excitation with 38 mCi of Cd-109.

6. Considerations on Polarized Beam FEA

Using polarized excitation, for a cost increase of approximately 50%, an automated FEA system can improve tenfold the assay time of low iodine concentration samples. Alternatively, for the same analysis time and accuracy, the lowest tracer working levels can be reduced by a factor of the square root of the improvement factor, in this case from ∿1 ppm to ∿0.3 ppm. It is apparent from Figure 24 that the use of more polarizing targets could yield even bigger improvements.

This level of improvement is not possible with either source or secondary fluorescence excitation. This can be seen from the fact that the photons from Am-241 are quite ideal for I excitation, and a secondary target could not improve substantially upon it. While more count rate is available, the effective rate could only go up by a factor of three to four. This can be seen as follows: at the upper concentration range of the system, I characteristic X-rays represent about 75% of the total counts being accumulated. For very low concentrations, where the characteristic X-ray contribution to total count rate is essentially zero, that count rate could be quadrupled at most.

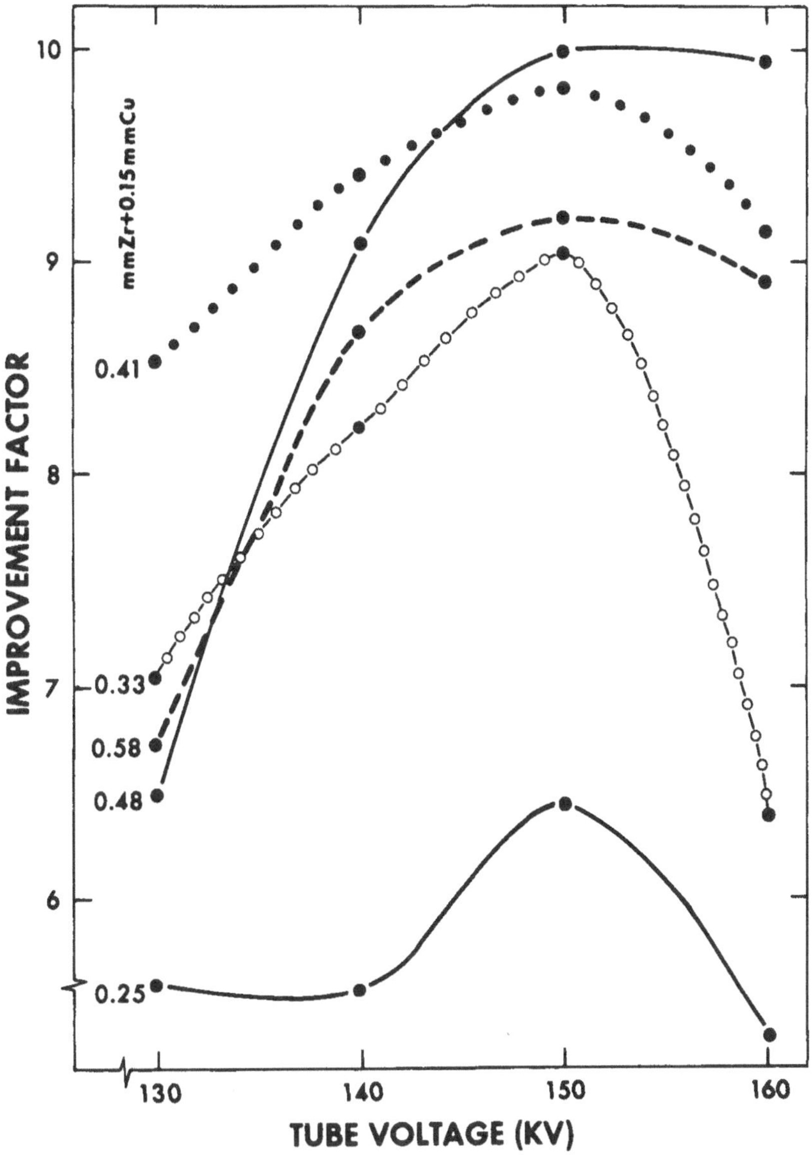

FIGURE 21. Improvement factor for FEA of iodine obtained with polarized excitation relative to Am-241 excitation, shown as a function of tube voltage for various filter thicknesses.

Given that the Compton-scattered X-rays have an average energy that is nearly twice as high as that of the characteristic X-rays, the pulsed optical reset will be activated more often, reducing the maximum allowable count-rate increase to less than a factor of four. This was corroborated with secondary tungsten and tantalum targets replacing the boron carbide polarizer. Because in the polarized excitation mode the characteristic X-rays represent a larger fraction of the total counts (Figure 25), if the system is operated near the upper count-rate range of the electronics, it will be more severely limited on dynamic range than the equivalent unpolarized excitation beam mode. This is not a serious limitation because in practice the advantages of the polarized mode are significant only for concentration levels of under 10 ppm (where counting times of 10 min or more are needed to obtain the desired accuracy).

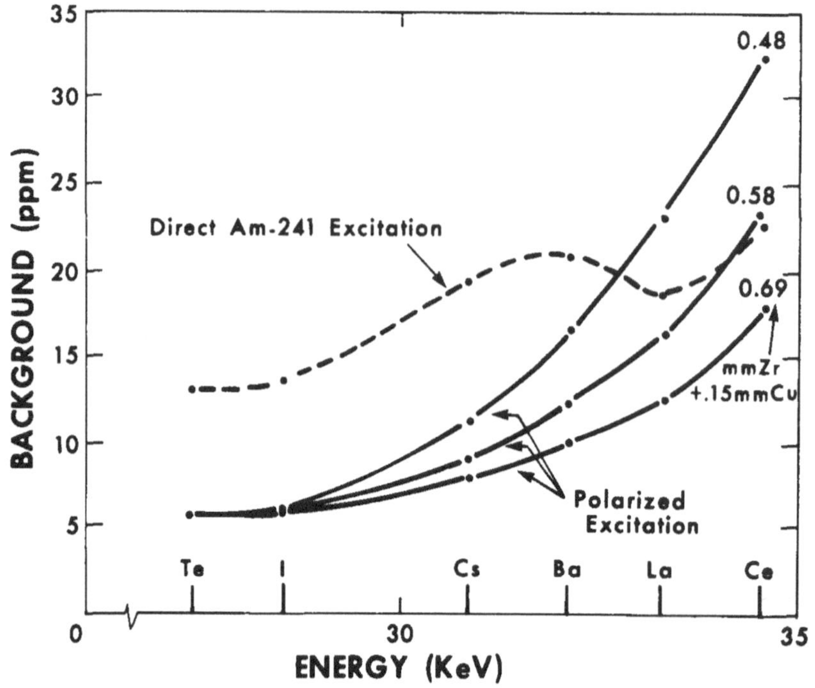

FIGURE 22. Background for various tracers of interest shown for polarized excitation with three filter thicknesses and for Am-241 excitation in the same detector. The bump for Cs and Ba is produced by scatter from the Am-241 33 keV line.

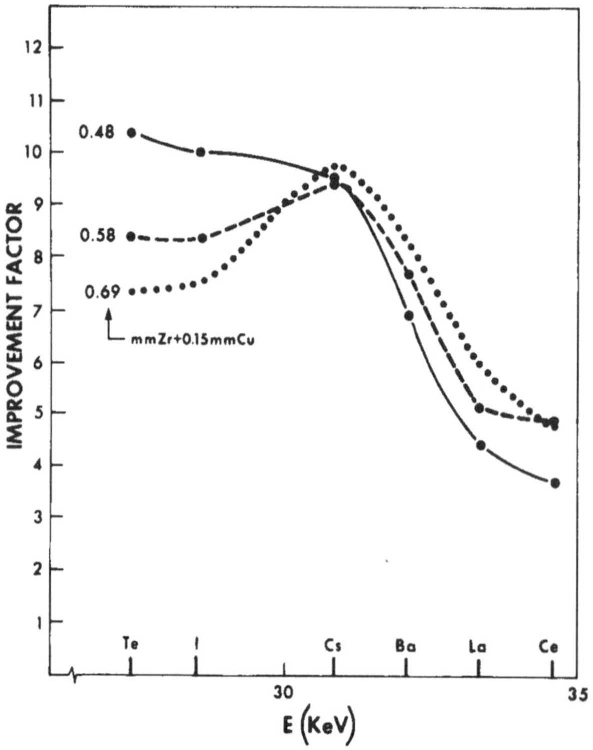

FIGURE 23. Improvement factor for polarized excitation of tracers of interest, using three different filter thicknesses and 150 kV tube voltage.

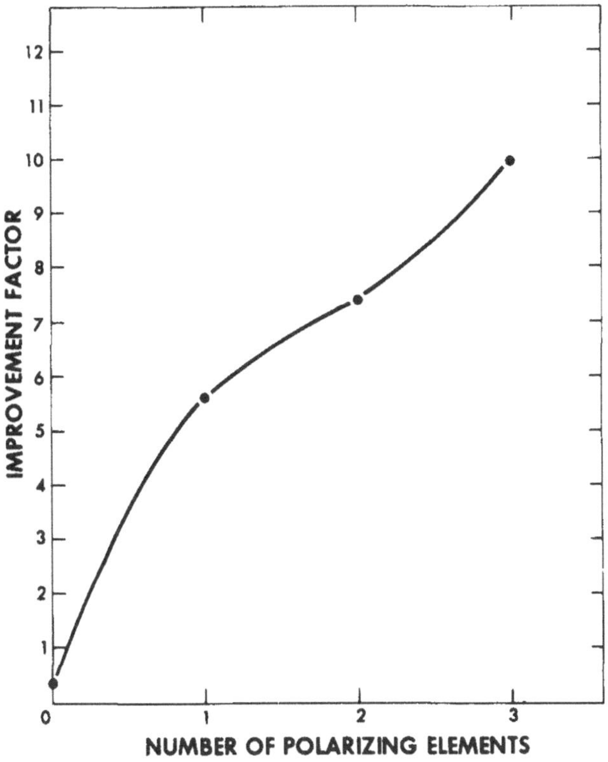

FIGURE 24. Improvement factor as a function of the number of polarizing targets. The value at zero is given by scattering from the chamber walls. It is apparent that further improvements may be possible by the addition of a fourth target.

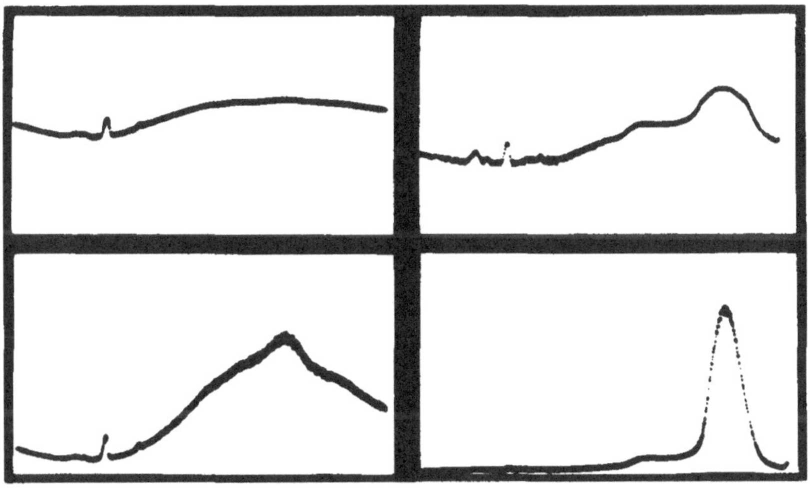

FIGURE 25. FEA spectra of a sample containing 14 ppmI, obtained with polarized excitation (right), shown in logarithmic (top) and linear (bottom) scales. The energy spanned along the horizontal axis ranges from 17 to 60 keV. Counting time was 600 sec in both modes.

Because an Am-241 source of adequate strength costs about 5% of the cost of an X-ray generator and tube, the measurement strategy should be to count all samples with more than 10 ppm using the source. Those samples that are known to contain lower tracer levels, or samples that the program finds (after say 1 min counting time) to contain less than 10 ppm, would be flagged for counting using the polarized beam.

Thus, high intensity sources of polarized X-rays for FEA are a cost-effective means for providing substantial reductions in assay time and/or minimum quantitable tracer levels. As experience is gained with this excitation mode, performance improvements surpassing those described here are certain to follow.

V. CONCLUSIONS

FEA is a tool which has usefulness in laboratory and clinical practice. In terms of accuracy and cost, it compares favorably to the techniques it replaces. Its main limitation is one of sensitivity. New techniques such as the use of polarized radiation for excitation and improved spectrometers, promise further improvements in sensitivity. More importantly, the use of physiologically compatible tracers is an important factor that is under continuous exploration.

Where applicable, FEA has been shown to be a powerful and rewarding tool.

REFERENCES

1. Kaufman, L., Deconinck, F., Price, D. C., Guesry, P., Wilson, D. C., Hruska, B., Swann, S. J., Camp, D. C., Voegele, A. L., Friesen, R. D., and Nelson, J. A., An automated fluorescent excitation analysis system for medical applications, *Invest. Radiol.*, 11, 210, 1976.
2. Camp, D. C., Voegele, A. L., Friesen, R. D., Kaufman, L., and Hruska, B., Automated sample changer for X-ray fluorescence analysis of bio-medical samples, *Chem. Inst.*, 7(2), 47, 1976.
3. Kaufman, L., Shosa, D., and Camp, D. C., A high intensity source of polarized X-rays for fluorescent excitation analysis, *IEEE Trans. Nucl. Sci.*, NS-24, 525, 1977.
4. Kaufman, L. and Camp, D. C., Polarized radiation for X-ray fluorescence analysis, *Adv. X-Ray Anal.*, 18, 247, 1975.

Chapter 6

FLUORESCENT EXCITATION ANALYSIS IN THE MEASUREMENT OF BODY WATER COMPARTMENTS, CELL SURVIVAL, AND SPECIFIC ORGAN FUNCTION

David C. Price

TABLE OF CONTENTS

I. Introduction 115

II. In Vitro Studies 115
A. Measurement of Body Compartments 115
1. Extracellular Fluid Space 116
2. Red Cell Volume 117
3. Plasma Volume 118
B. Blood Cell Survival Studies 120
C. Glomerular Filtration Rate 121
D. Other In Vitro Applications 122

III. In Vivo Studies 122
A. Cardiac Output 122
B. Cerebral Blood Flow and Blood Volume 124
C. Clearance of Heavy Metal Powders from the Lungs 124
D. Other In Vivo Applications 125

IV. Conclusions 126

References 126

I. INTRODUCTION

Radiotracers were in use for evaluation of various body water compartments,[1-4] survival of circulating blood cells (particularly erythrocytes),[9-11] and specific organ function (e.g., thyroid and bone marrow)[5-8] long before the development of scintigraphic imaging in nuclear medicine. Fluorescent excitation analysis (FEA) of nonradioactive tracers has similarly been applied to these three types of tracer studies in its early development. Imaging by FEA has so far been limited to the thyroid, the only organ in the body having sufficient intrinsic concentration of a medium-atomic-weight element (iodine) to permit fluorescent imaging (see Chapter 9). The present chapter reviews a number of in vitro and nonimaging in vivo applications of FEA as they have evolved to date. Some applications of FEA to hepatic evaluation are covered in Chapter 7 on radiographic contrast agents.

II. IN VITRO STUDIES

A. Measurement of Body Compartments

As with the use of radioactive tracers, the dilutional assay principle can be successfully applied to FEA of stable tracers in the measurement of various body water compartments. There are two important requirements in the development of such techniques:

1. The availability of elemental tracers having relatively high atomic numbers which are appropriate for labeling the compartment of interest, and
2. A tracer methodology which does not disturb the physiology of the compartment under study.

This latter criterion is particularly important in that relatively large amounts of stable tracer are necessary for statistically valid FEA, introducing the possibility of pharmacologic or even toxic effects from the tracer.

1. Extracellular Fluid Space

Bromine both in stable[12] and radioactive[13] form has been a traditional tracer for measurement of the extracellular fluid volume (ECFV), functioning as a chemical analogue of extracellular chloride. Other ECFV tracers such as thiocyanate,[14] thiosulfate,[2] sucrose,[13] etc. have also been used, although none has demonstrated clear superiority over bromine. Since FEA of stable bromine is relatively simple, it appeared logical to first explore bromide as the tracer for determination of the ECFV by FEA. Studies in animals, then in children[15] and adults[16] have confirmed the validity of the technique in comparison with both the chemical and the 82-Bromine methods presently in wide use (Tables 1 and 2). In contrast to the chemical method, FEA provides higher statistical accuracy with much shorter sample assay times, permitting many more studies per technologist per working day.[15] In contrast to the radio-bromide method, FEA eliminates radiation exposure to the subject and to the attending personnel. In addition, there is a long shelf-life and a considerably lower cost for the stable tracer.

As a standardized procedure, 40 mEq of sodium bromide (0.57 mEq/kg in children) is administered orally or intravenously in the resting state after an overnight fast. Two anticoagulated blood samples of 3 to 5 cc are drawn at 6 and 6¼ hr, the plasma separated, and the bromine concentrations assayed in 1 to 2 cc plasma aliquots. Cadmium-109 is used as the excitation source, resulting in the production of 11.91 and 13.30 kev K-alpha and K-beta characteristic X-rays of bromine. An aliquot of the administered dose is retained and assayed at the same time. Of particular note is the fact that FEA is linear over several orders of magnitude of tracer concentration, eliminating the need for dilution of the retained standard before counting. Urine assay for bromine loss durng the 6¼ hr equilibration is helpful, but not essential, as the tracer

TABLE 1

Comparison of Fluorescent and Chemical Determinations

	Corrected bromide space (*l*)		
	Fluorescent	Chemical	Difference (%)
Children			
1	3.45	3.62	4.7
2	13.12	11.63	−11.3
3	5.70	5.60	−1.7
4	9.45	9.08	−3.9
5	6.28	5.90	−6.0
6	6.59	7.28	10.5
Rats			
1	0.121	0.127	5.4
2	0.136	0.141	3.5
Mean			0.2

TABLE 2

Comparative Extracellular Fluid Spaces in Adults

Patient	Sex	Age (Years)	Weight (k)	Diagnosis	^{82}Br ECF (*l*)	^{82}Br % Body Weight	^{80}Br-FEA ECF (*l*)	^{80}Br-FEA % Body Weight
1	M	18	72	Urticaria	17.3	24.0	18.3	25.5
2	M	23	62	Asthma	19.7	31.8	22.2	35.8
3	M	24	87	Obesity	21.8	25.1	20.1	23.1
4	M	42	59	Liver disease	19.1	32.4	18.8	31.8
5	M	29	68	Obesity	14.5	21.3	14.0	20.5
6	F	71	52	Myocardial infarction	13.9	26.7	13.8	26.5
7	F	46	52	Hypertension	18.2	35.1	15.3	29.4
8	F	49	56	Cirrhosis, ascites	22.1	39.4	21.1	37.7
9	M	69	66	Hypertension, CHF	20.0	30.3	18.9	28.7
10	M	72	74	Pneumonia	19.5	26.3	17.7	23.9
11	M	39	49	Sickle cell disease	19.3	39.4	16.2	33.0
				Mean		30.2		28.7
				S.D.		5.8		5.2

loss during this period is customarily only 0.4 to 1.2% (average 0.7%) of the administered dose. It is necessary to introduce mathematical corrections for the Gibbs-Donnan effect and plasma protein content (0.93), as well as correction for 5.7% average tracer diffusion into red blood cells. The ECFV is then determined from the following formula:

$$\text{ECFV (1)} = \frac{(\text{Total tracer introduced} \times 0.943) - (\text{Urinary tracer loss})}{(\text{Average concentration of Br}/\ell \text{ plasma at 6 hr})} \times 0.93$$

Table 2 summarizes the data obtained in a comparative study of ECFV in 11 patients utilizing the FEA and the 82-Br methods. The validity of this method is clearly evident.

2. Red Cell Volume

Development of an FEA method for safe and accurate determination of red cell volume in man has been complicated by difficulties in finding an appropriate stable tracer. Although radioactive chromium is a customary tracer for red cell labeling in clinical studies, the amount of stable chromium necessary for effective FEA after dilution in total blood volume would be well above the levels of chromium known to be toxic to the erythrocytes (50 μg Cr/m*l* erythrocytes).[17] Initial attempts in the laboratory to label erythrocytes with stable cesium as a potassium analogue were unsuccessful. Only 1 to 2% of the cesium diffuses spontaneously into the erythrocytes in spite of incubation periods as long as 24 hr. Apparently the cesium ion does not freely traverse the erythrocyte membrane, and is too large to fit into the sodium-potassium pump mechanism.

It has been known for several years that polyene antibiotics such as Nystatin and Amphotericin B can create reversible pores or channels in the erythrocyte membrane, which permit free diffusion of monovalent ions.[18,19] Incubation of erythrocytes in isotonic cesium chloride (2.6%) with addition of 50 μg/m*l* Nystatin results in cellular incorporation of a large proportion of the cesium.[20] The Nystatin is removed by an isotonic cesium chloride wash, without evident injury to the red cells. In a comparative study of 11 humans, the cesium-FEA red cell volume did not differ significantly from the traditional 51-Chromium technique (Table 3).[20] It was noted, however, that red cell cesium disappeared with a relatively rapid half-time of 2.4 days compared to the 51-Chromium red cell half-time of 25.2 days (Figure 1). This accelerated cesium disappearance may be due to leakage of cesium out of the labeled cells, since damage to the cells by the labeling process itself was excluded by double labeling of the same cell preparation. The cesium-FEA method has also recently been applied to the experimental evaluation of massive acute hemorrhage in newborn lambs.[21]

The technique for determination of red cell volume by FEA of cesium-labeled autologous erythrocytes is as follows:

a. Draw 15 m*l* patient blood under sterile conditions into a heparinized syringe. Transfer to a sterile, pyrogen-free, sealed tube.
b. Centrifuge rapidly for 15 min and remove the supernatant plasma.
c. To the packed red blood cells, add an equal volume of sterile, pyrogen-free 2.6% cesium chloride (pH 7.4) containing Nystatin at a concentration of 50 μg/m*l*.[20]
d. Incubate for 30 min at room temperature after mixing well. Agitate periodically.
e. Centrifuge, remove the supernatant, and wash once with 2.6% cesium chloride solution without Nystatin.
f. Wash 5 to 6 times further with sterile normal saline, and reconstitute terminally with sterile normal saline to approximately a 50% hematocrit.
g. Inject intravenously a weighed amount of the labeled red cells, retaining a 1.5 to 2 m*l* aliquot as a standard.
h. Draw 2 to 3 m*l* anticoagulated blood samples at 10 min and at 40 min for hematocrit determination and for cesium assay by FEA.
i. The red cell volume (RCV) is calculated by the following formula:

$$\text{RCV (l)} = \frac{(\text{Weight of injectate}) \times (\text{Cs concentration of injectate}) \times (\text{fractional hematocrit})}{(\text{Specific gravity of blood}) \times (\text{cesium concentration in blood} - \text{blood cesium background})}$$

j. Specific gravities for bloods of varying hematocrit are given in Table 4. The 40-min blood sample is customarily used for calculation. If there has been a significant drop in cesium concentration from 10 to 40 min, this may reflect unusual leakage of cesium out of the erythrocytes. In this case the 10-min value should be used.

Cesareo[22] has indicated the intention to explore rubidium, tin, and nickel compounds for labeling of erythrocytes in the determination of red cell volume by FEA. To date, no results of such studies have been published.

3. Plasma Volume

The traditional label for determination of plasma volume has been radio-iodinated human serum albumin.[23] Although no studies have been reported utilizing fluorescible tracers for this purpose, we have begun to explore the use of heavily iodinated human serum albumin. Electrolytic iodination of albumin at the level of 0.5 to 1 iodine atoms per albumin molecule results in a good tracer for animal studies, although very large amounts of intravenous albumin would then have to be utilized in humans. Table 5 indicates the good correlation obtained in rabbits between FEA of such an iodine-labeled albumin preparation and ^{125}I-albumin in the measurement of plasma volume. Exploration of the technique for heavy iodination of albumin (15 to 20 iodine atoms per albumin molecule) is presently under way in our laboratory for future human application.

Cesareo[24] has indicated the feasibility of determining plasma volume (and thus total blood volume) by back-extrapolation of the second exponential component of the dis-

TABLE 3

Human Red Cell Volumes

Time after injection	10 min			40 min		
		Cs (FEA)			Cs (FEA)	
Tracer	^{51}Cr Vol. (*l*)	Vol. (*l*)	Diff. (%)	^{51}Cr Vol. (*l*)	Vol. (*l*)	Diff. (%)
1	2.30	2.44	6.1	2.24	2.15	−4.0
2	1.69	1.64	−3.0	1.71	1.73	1.2
3	0.75	0.71	−5.3	0.76	0.78	2.6
4	1.39	1.45	4.3	1.41	1.46	3.5
5	1.10	1.16	5.5	1.08	1.33	23.1
6	1.35	1.38	2.2	1.35	1.35	0.0
7	1.01	0.98	−3.0	1.01	1.00	−1.0
8	3.46	3.26	−5.8	3.56	3.41	−4.2
9	2.29	2.22	−3.1	2.22	2.24	0.9
10	1.24	1.29	4.0	1.28	1.36	6.3
11	2.91	2.93	0.7	2.99	2.96	−1.0
Mean			0.2			2.5
Standard Deviation			4.4			7.5
p			<.001			<.001

CESIUM AND ^{51}CHROMIUM BLOOD DISAPPEARANCE (Human)

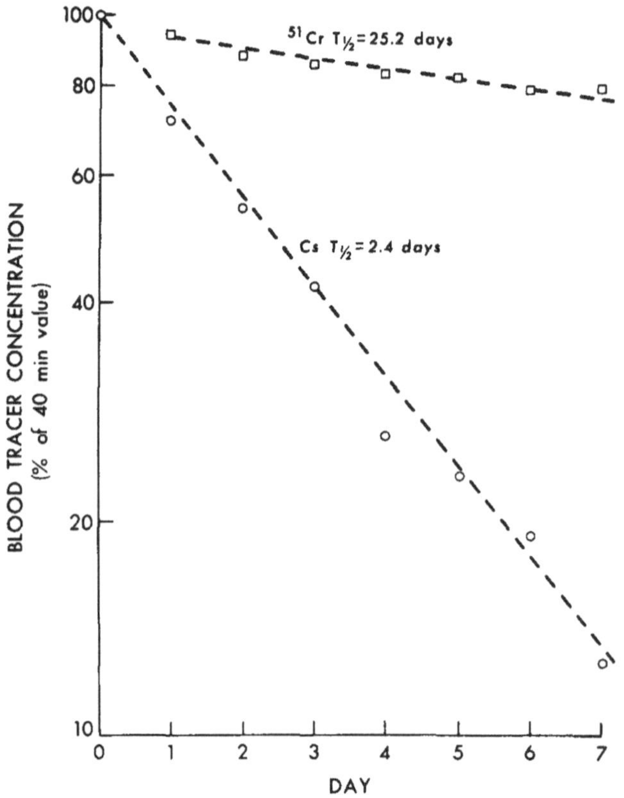

FIGURE 1. Comparison between ^{51}Cr red cell survival half-time (25.2 days) and cesium-red cell disappearance half-time (2.4 days) in a normal adult, consistent with cesium elution from the labeled erythrocytes.

TABLE 4

Whole Blood Specific Gravities

Hematocrit (%)	Specific gravity	Hematocrit (%)	Specific gravity
21.0	1.0421	36.0	1.0519
22.0	1.0428	37.0	1.0525
23.0	1.0434	38.0	1.0532
24.0	1.0441	39.0	1.0538
25.0	1.0447	40.0	1.0545
26.0	1.0454	41.0	1.0551
27.0	1.0460	42.0	1.0558
28.0	1.0467	43.0	1.0565
29.0	1.0473	44.0	1.0571
30.0	1.0480	45.0	1.0577
31.0	1.0486	46.0	1.0584
32.0	1.0493	47.0	1.0590
33.0	1.0499	48.0	1.0597
34.0	1.0506	49.0	1.0603
35.0	1.0512	50.0	1.0610

TABLE 5

Rabbit Plasma Volumes by FEA

^{125}I-Albumin	I-Albumin (FEA)	Difference (%)
94.9 cc	99.3 cc	+4.6
153.5	140.0	−8.8
134.1	140.4	+4.7
132.7	132.6	−0.1
133.6	144.1	+7.9
141.4	151.0	+6.8
146.5	148.5	+1.4
140.3	121.6	−13.3
Total		$+0.4 \pm 7.6$

appearance curve of an angiographic dye from the blood. No confirmatory correlative study was carried out, and the statistical problems in back-extrapolation of a complex, rapid disappearance curve would appear to be considerable. Cesareo[22] has also indicated the intent to explore FEA of iodinated serum components in the measurement of plasma volume.

B. Blood Cell Survival Studies

The application of FEA tracers to cell survival studies is still almost totally unexplored. We have noted that the disappearance of cesium from labeled red blood cells was extremely rapid (half-time 2.4 days in one human, 31.5 ± 2 hr in four rabbits), much more rapid than the radio-chromium survival half-times.[20] It was concluded that cesium red cell labeling by this technique could not be utilized for cell survival studies, due perhaps to leakage of cesium out of the erythrocytes in circulation.

Cesareo[25,26] has demonstrated effective techniques for fluorescence of selenocystine-labeled and rubidium-labeled human platelets, including one normal human study and one study of shortened platelet survival due to EDTA damage to the platelets. The method appears effective, although more complete documentation of its accuracy in comparison to the standard 51-Chromium or DF^{32}P technique is needed.

C. Glomerular Filtration Rate

One of the traditional methods for estimation of glomerular filtration rate (GFR) has been measurement of the blood clearance of labeled iothalamate by the kidneys.[27,28] This method gives data comparable to the figures derived from inulin clearance. It would seem logical that clearance of nonradioactive iothalamate would give an equally valid estimate of GFR. This has proven to be so in animals and in humans.[29]

Guesry et al.[29] compared FEA of iothalamate both with inulin clearance in constant infusion studies and with ^{125}I-iothalamate in single injection studies, as well as evaluating two-exponential and single-exponential models for the derivation of GFR (Figure 2). As a result, their simple and accurate technique requires only three blood samples at 2, 3, and 4 hr after injection of a single dose of nonradioactive iothalamate intravenously. Alazraki et al.[30] have demonstrated good correlation in dogs between the iothalamate-FEA technique and 14-Carbon inulin clearance, both by constant infusion and by subcutaneous injection of the tracer. No evaluation was made in their study of the single-intravenous-injection method.

As recommended by Guesry et al.,[29] a standardized method for GFR determination by FEA is as follows:

a. Draw a preinjection blood sample ino EDTA and centrifuge to obtain the background plasma sample.

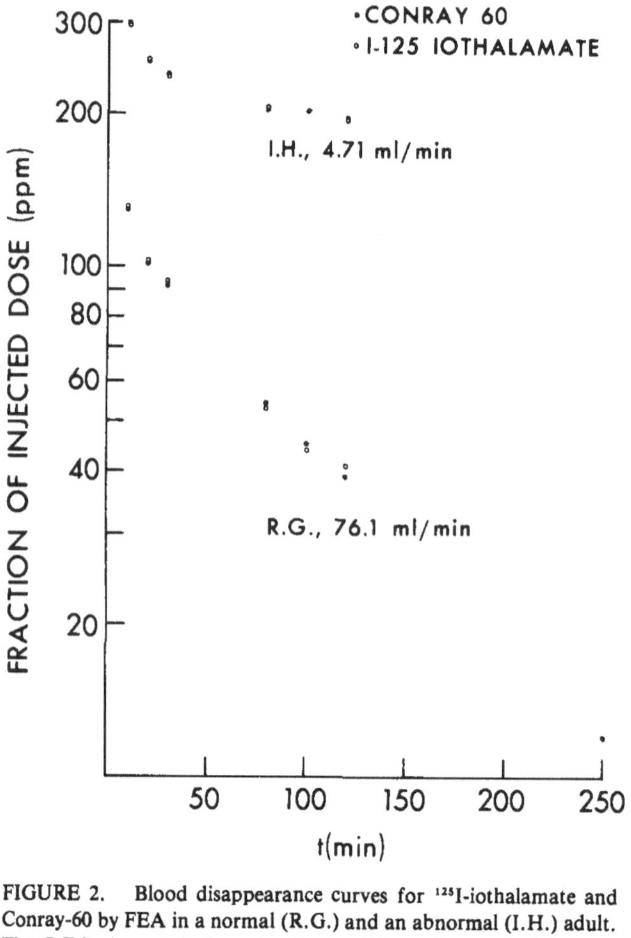

FIGURE 2. Blood disappearance curves for ^{125}I-iothalamate and Conray-60 by FEA in a normal (R.G.) and an abnormal (I.H.) adult. The G.F.R. in each instance is calculated from the terminal slope as described in the text.

b. Inject 0.15 mℓ/kg of Conray-60 intravenously, weighing the syringe before and after injection to assure accurate determination of the amount of injectate. When the infusion is made through a catheter or "butterfly", the catheter must be thoroughly cleared intravenously with normal saline. The Conray-60 should not be injected if the patient is known to be allergic to iodinated contrast media.
c. Anticoagulated blood samples (4 to 5 mℓ) are drawn at 2, 3, and 4 hr into EDTA. Plasma is separated by centrifugation.
d. Iodine concentrations are measured by FEA on an aliquot of the injectate as well as the background and 2-, 3-, and 4-hr samples. The Conray-60 iodine concentration has been found to be quite stable at 282 mg iodine/mℓ, with less than 1% free iodine if properly stored. Thus, assay of the injectate with each study is not really necessary. Conray-60 has a specific gravity of 1.312.
e. The half-time clearance of plasma iodine (after background subtraction) is calculated by single-exponential analysis of the three data points, and the GFR is calculated as follows:

$$\text{GFR (m}\ell\text{/min)} = \frac{\left(\text{Weight of injectate} \times \frac{282}{1.312}\right) \times 0.693}{(T_0 \text{ intercept}) \times T_{1/2}\ (\text{min}) \times 1.030^{*}}$$

*Specific gravity of serum

D. Other In Vitro Applications

The radiolabeled microsphere technique described by Rudolph and Heyman[31] in 1967 has become a very widely utilized method for evaluating changes in relative organ or regional blood flow in animals under various pathophysiologic conditions. Limitations in the technique include the inability to utilize more than five gamma-emitting microsphere tracers at one time (due to the energy resolution limitations of sodium iodide detectors), the considerable cost of such tracers (due in part to the costs of radionuclide production), and a limited shelf life which is determined by the physical half-life of each radionuclide. Quarantine and disposal of radioactive animals present additional difficulties with the technique. Payne and co-workers[32,33] of the same group have begun to explore FEA of microspheres labeled with stable elemental tracers for this purpose, and have had considerable success to date using silver, barium, cesium, and lanthanum in simultaneous determinations of relative myocardial blood flow (Figure 3). Further exploration of this new methodology is possible and is presently under way.

III. IN VIVO STUDIES

A. Cardiac Output

Two groups have published studies confirming the feasibility of calculating cardiac output by FEA of an intravenous angiographic dye injection. Kaufman et al.[34] demonstrated in dogs that in vivo FEA of a 5 mℓ bolus of iothalamate (Conray-60) during first-pass cardiac kinetics permitted accurate measurement of cardiac output, utilizing a modification of the Stewart-Hamilton dye dilution formula for the calculation. Figure 4 illustrates a typical cardiac curve obtained with 0.2-sec counting intervals. Comparison with the indocyanine green technique in four dogs revealed good correlation between the two methods. The excitation source utilized was a 95 keV three-phase rotating-tube X-ray generator.

Cesareo et al.[24] injected 1 mℓ Angiografin (containing 305.8 mg iodine) into the femoral vein of rabbits, sampling arterial dye concentrations during the first cardiac pass by a novel technique. A continuous, fine spurt of femoral artery blood was di-

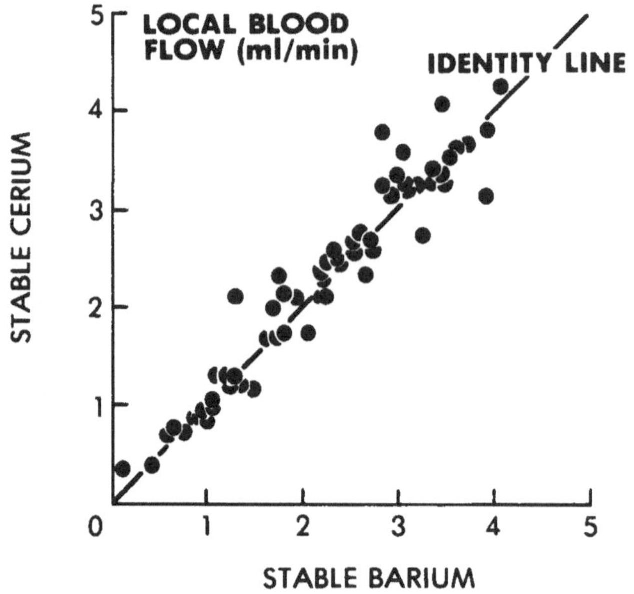

FIGURE 3. Comparison of local blood flow in various dog tissues, determined by simultaneous injection and measurement of stable barium- and cerium-loaded microspheres, assayed by FEA.

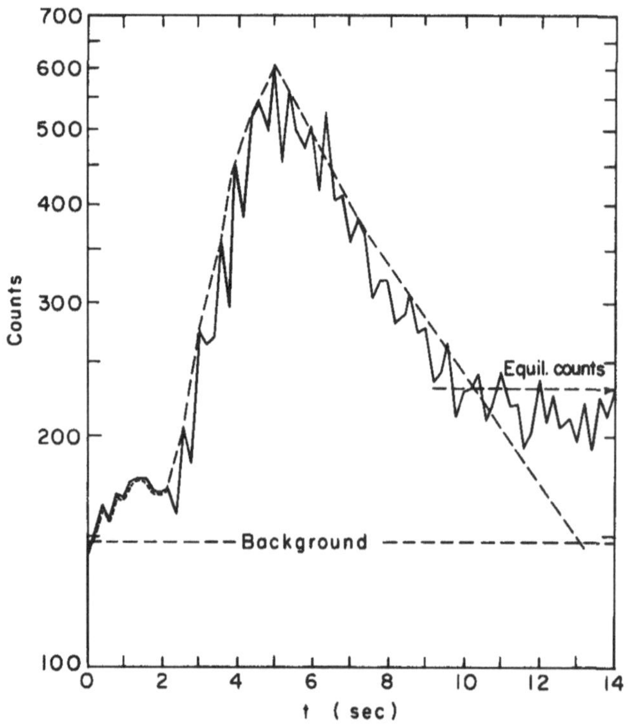

FIGURE 4. Cardiac output curve derived from 0.2-sec external excitation probe counts over the precordium of a dog following intravenous injection of a 5 m*l* bolus of Conray-60 (iothalamate).

rected onto a filter paper ribbon moving at fixed speed. FEA of the ribbon then demonstrated along its length a curve of fluorescible iodine concentrations which indicated the resulting first-pass arterial concentration curve. An annular 238-Plutonium source with a xenon-methane proportional counter made up the detection system. Comparison with the thermal dilution method confirmed the effectiveness of the technique in seven rabbits. Blood volume and circulation time were additional pieces of cardiovascular information derived from this study.

Neither technique has yet been applied to the measurement of cardiac output in humans. Radiation exposure by direct thoracic FEA is quite low, in the order of 15 to 30 mr localized to the thorax. The necessity for direct and continuous arterial sampling in Cesareo's method would be a significant disadvantage for routine human studies.

B. Cerebral Blood Flow and Blood Volume

Hoffer et al.[35] have indicated the feasibility in two monkeys of deriving cerebral blood flow data by in vivo FEA. In one instance, Renografin-76 (meglumine diatrizoate) was injected intraarterially and its cerebral transit time monitored by an external FEA probe system. In the other, significant intracerebral xenon concentrations were achieved by respiration over 6 min of an 80% xenon/20% oxygen mixture. Cerebral wash-out was then calculated during the monophasic clearance curve. Although both techniques were later tried in humans,[36] no specific calculations of cerebral blood flow nor comparisons with standard methods have yet been reported.

Much more extensive work has been performed by Ter-Pogossian et al.[37-40] on the measurement of regional cerebral blood volume in dogs, monkeys, and man by in vivo FEA. A collimated X-ray tube source and Si(Li) detector permitted evaluation of a 1 cc volume of the brain. Comparisons were made with the intracerebral content of 51-Chromium-tagged autologous red blood cells, with ± 8% correlation. In vivo human cerebral blood volume was noted to be 3.20 ± 0.94 cc/100 g tissue in the posterior frontal lobe. Studies in monkeys demonstrated an inverse linear relationship between arterial blood pressure and regional cerebral blood volume.

The technique is a very interesting one which awaits further human application.

C. Clearance of Heavy Metal Powders from the Lungs

Both insufflated tantalum and barium sulfate powders have been used for radiologic delineation of pulmonary airways.[41] The rate of clearance of such powders from the lungs has been studied both in the evaluation of tantalum as a radiologic contrast agent, and in the documentation of disturbed pulmonary clearance mechanisms in such diseases as bronchiectasis, emphysema, chronic obstructive pulmonary disease, and bronchogenic carcinoma.[41]

However, the radiographic techniques available for measurement of pulmonary clearance rates of metallic powders are limited in usefulness by their relative insensitivity, as well as by variations in X-ray exposure and in film response. The secondary phase of pulmonary dispersion of such powders is impossible to monitor by simple X-rays. Tantalum-182 is available as a radioisotopic tracer to follow pulmonary tantalum clearance, but its physical and biological half-times are long, precluding human use and complicating its use in animal studies.

FEA of both tantalum and barium is technically easy to perform, and readily modified for in vivo applications. Kaufman and Gamsu[43] have demonstrated in animals and in man both the feasibility and utility of this approach to pulmonary clearance studies. Localized regions of lung may be accurately monitored, and radiation exposure to the subject is limited to the time of actual measurement and to the region of interest. Nonradioactive tantalum represents a benign and convenient tracer for studies in man.

FEA of tantalum has been performed using the 97 and 103 keV gamma rays of 2 Ci of gadolinium-135 as the excitation source.[43] Since gadolinium decays to europium, the K X-rays of the latter are then useful for the simultaneous excitation of barium. A wide-field-of-view collimator permits assay of a 10 cc volume, thus minimizing variability in measurements due to slight shifts in position. Up to 2 cm displacement in phantoms produced no significant change in the assay value.

Studies to date have demonstrated that although there is predominant clearing of radiographically-evident tantalum in the first 2 to 3 days, FEA demonstrates a significant residual fraction of the tracer having a very slow background clearance rate (Figure 5). This would appear to be more diffusely dispersed bronchiolar and alveolar tantalum whose slow biological turnover has been completely overlooked by the earlier radiographic methods.

D. Other in Vivo Applications

Hermann and Kiker[44] have indicated a novel method for monitoring changes in intracranial pressure. A two-chamber reservoir inserted on either side of the calvarium contains a cesium nitrate solution. Any increase in intracranial pressure shifts an increased volume of fluid from the internal to the external chamber, where FEA accurately quantitates that volume. No studies have yet been reported in humans.

An unusual technique for in vivo FEA of gaseous xenon in the lateral ventricles of monkeys and of man has been reported by Kaufman et al.[45] Accurate measurement of the rate of systemic absorption of xenon from lateral ventricles over a number of hours was easily achieved. Of particular interest in these preliminary studies was the fact that measurable xenon cleared more rapidly from the ventricles than did radiographically evident gas. This fact suggests that there is diffusion of circulating gases from the intracranial blood into the xenon-filled lateral ventricles, attempting to reach

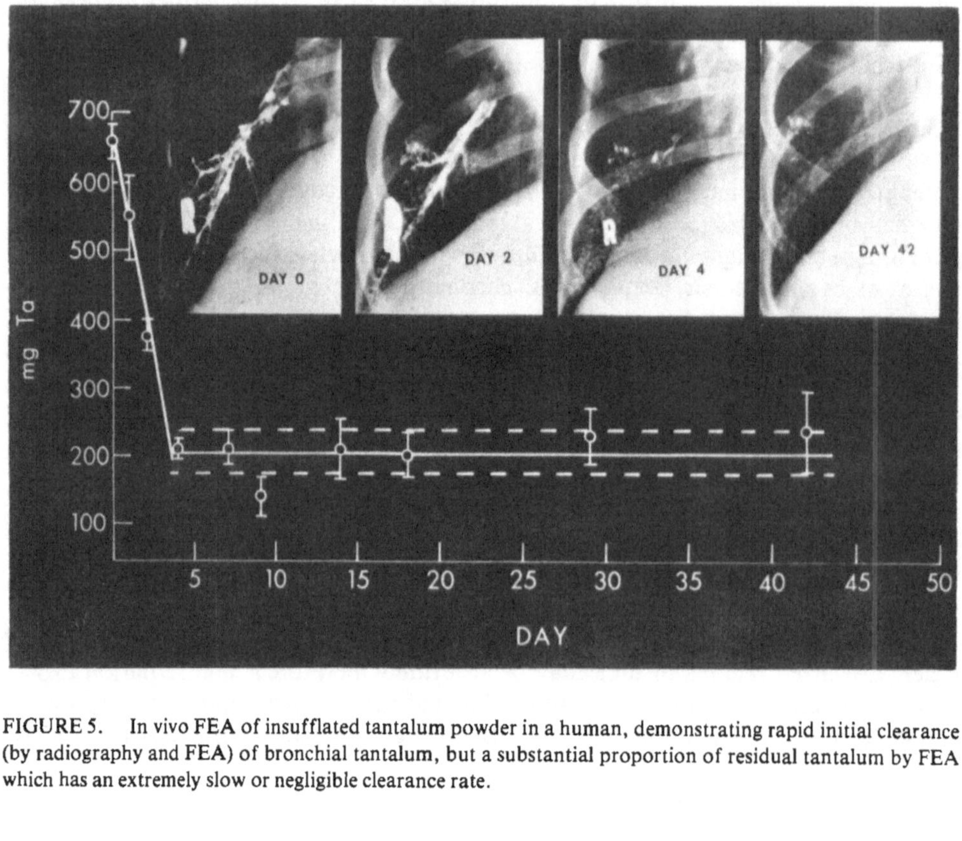

FIGURE 5. In vivo FEA of insufflated tantalum powder in a human, demonstrating rapid initial clearance (by radiography and FEA) of bronchial tantalum, but a substantial proportion of residual tantalum by FEA which has an extremely slow or negligible clearance rate.

gaseous equilibrium with the newly-created space in the ventricles. The technique of introducing xenon or air into the lateral ventricles by lumbar subarachnoid injection has been used in the past to monitor rate of disappearance of gas from the lateral ventricles, and thus infer the rate of cerebrospinal fluid formation.[46,47] The FEA findings with xenon demonstrate that this postulate is invalid, and that the rate of clearance of the radiographically-measured ventricular gas is not a simple function of the rate of cerebral spinal fluid formation and absorption.

IV. CONCLUSION

Exploration of FEA in the study of body water compartments, cell survival, and specific organ function is resulting in a number of new and useful measurement techniques. Various chemical innovations, such as the use of polyene antibiotics to facilitate the labeling of erythrocytes, will make many more types of physiologic tracer studies adaptable to the FEA technique. This will bring the numerous advantages of stable tracers to human studies, such as reduced or absent radiation exposure of the patient and attending personnel, inexpensive tracers with long shelf lives, new types of physiologic tracer studies, etc. The applications reviewed in this chapter will prove to be only the beginning of a large number of new biomedical FEA techniques yet to be developed.

REFERENCES

1. Gregersen, M. I. and Rawson, R. A., Blood volume, *Physiol. Rev.*, 39, 307, 1959.
2. Edelman, I. S., Olney, J. M., James, A. H., Brooks, L., and Moore, F. D., Body composition: studies in the human being by the dilution principle, *Science*, 115, 447, 1952.
3. Cheek, D. B., Extracellular volume: its structure and measurement and the influence of age and disease, *J. Pediatr.*, 58, 103, 1961.
4. Pace, N., Kline, L., Schachman, H. K., and Harfenist, M., Studies on body composition. IV. Use of radioactive hydrogen for measurement *in vivo* of total body water, *J. Biol. Chem.*, 168, 459, 1947.
5. Hertz, S., Roberts, A., and Evans, R. D., Radioactive iodine as an indicator in the study of thyroid physiology, *Proc. Soc. Exp. Biol. Med.*, 38, 510, 1938.
6. Hamilton, J. G. and Soley, M. H., Studies in iodine metabolism by the use of a new radioactive isotope of iodine, *Am. J. Physiol.*, 127, 557, 1939.
7. Huff, R. L., Hennessy, T. G., Austin, R. E., Garcia, J. F., Roberts, B. M., and Lawrence, J. H., Plasma and red cell iron turnover in normal subjects and in patients having various hematopoietic disorders, *J. Clin. Invest.*, 29, 1041, 1950.
8. Pollycove, M. and Mortimer, R., The quantitative determination of iron kinetics and hemoglobin synthesis in human subjects, *J. Clin. Invest.*, 40, 753, 1961.
9. Ebaugh, F. G., Jr., Ross, J. F., and Emerson, C. P., The use of chromium-51 as an erythrocyte-tagging agent for determining the survival of transfused and autotransfused erythrocytes, *J. Clin. Invest.*, 32, 563 and 1260, 1953.
10. Cohen, J. A. and Warringa, M. G. P. J., The fate of P-32 labeled diisopropylfluorophosphate in the human body and its use as a labeling agent in the study of the turnover of blood plasma and red cells, *J. Clin. Invest.*, 33, 459, 1954.
11. Morgan, M. C., Keating, R. P., and Reisner, E. H., Jr., Survival of radiochromate-labeled platelets in rabbits, *J. Lab. Clin. Med.*, 46, 521, 1955.
12. Brodie, B. B., Brand, E., and Leshin, S., The use of bromide as a measurement of extracellular fluid, *J. Biol. Chem.*, 130, 155, 1939.
13. Gamble, J. L., Jr., Robertson, J. S., Hannigan, C. A., Foster, C. G., and Farr, J. E., Chloride, bromide, sodium and sucrose spaces in man, *J. Clin. Invest.*, 32, 483, 1953.
14. Macaulay, D., Dye space and thiocyanate space estimation in children, *Clin. Sci.*, 13, 345, 1954.
15. Kaufman, L. and Wilson, C. J., Determination of extracellular fluid volume by fluorescence excitation analysis of bromine, *J. Nucl. Med.*, 14, 812, 1973.

16. Price, D. C., Kaufman, L., and Pierson, R. N., Jr., Determination of the bromide space in man by fluorescent excitation analysis of oral bromine, *J. Nucl. Med.*, 16, 814, 1975.
17. Cline, M. J. and Berlin, N. I., The red cell chromium elution rate in patients with some hematologic disorders, *Blood*, 21, 63, 1963.
18. Butler, W. T., Alling, D. W., and Cotlove, E., Potassium loss from human erythrocytes exposed to Amphotericin B, *Proc. Soc. Exp. Biol. Med.*, 118, 297, 1965.
19. Finkelstein, A. and Holz, R., Aqueous pores created in thin lipid membranes by the polyene antibiotics Nystatin and Amphotericin B, *Membranes*, 2, 377, 1973.
20. Price, D. C., Swann, S. J., Hung, S. T.-C., Kaufman, L., Huberty, J. P., and Shohet, S. B., The measurement of circulating red cell volume using non-radioactive cesium and fluorescent excitation analysis, *J. Lab. Clin. Med.*, 87, 535, 1976.
21. Spitzer, A. R., Phibbs, R. H., and Schlueter, M., Massive hemorrhage in the newborn lamb, *Pediatr. Res.*, 12, 1027, 1978.
22. Cesareo, R., Trace elements analysis in biological samples by radioisotopic X-ray fluorescence, *Eur. J. Nucl. Med.*, 1, 49, 1976.
23. Crispell, K. R., Porter, B., and Nieset, R. T., Studies of plasma volume using human serum albumin tagged with 131-Iodine, *J. Clin. Invest.*, 29, 25, 1953.
24. Cesareo, R., Tallarida, G., and Baldoni, F., Determination of hemodynamic parameters in the rabbit by X-ray fluorescence excitation, *Int. J. Appl. Radiat. Isot.*, 26, 285, 1975.
25. Cesareo, R., del Principe, D., Mancuso, G., and Tallarida, B. M., In vitro labelling of platelets with stable selenocystine, *Int. J. Appl. Radiat. Isotop.*, 27, 324, 1976.
26. Cesareo, R., del Principe, D., Tallarida, B., and Mancuso, G., In vitro labelling of platelets with stable rubidium compounds, *Eur. J. Nucl. Med.*, 3, 91, 1978.
27. Sigman, E. M., Elwood, C. M., and Knox, F., The measurement of glomerular filtration rate in man with sodium iothalamate 131-I (Conray), *J. Nucl. Med.*, 7, 60, 1966.
28. Maher, F. T. and Tauxe, W. N., Renal clearance in man of pharmaceuticals containing radioactive iodine, *JAMA*, 207, 97, 1969.
29. Guesry, P., Kaufman, L., Orloff, S., Nelson, J. A., Swann, S., and Holliday, M., Measurement of glomerular filtration rate by fluorescent excitation of non-radioactive meglumine iothalamate, *Clin. Nephrol.*, 3, 134, 1975.
30. Alazraki, N., Verba, J. W., Henry, J. E., Becker, R., Taylor, A., Jr., and Halpern, S. E., Noninvasive determination of glomerular filtration rate using X-ray fluorescence, *Radiology*, 122, 183, 1977.
31. Rudolph, A. M. and Heymann, M. A., The circulation of the fetus in utero: methods for studying distribution of blood flow, cardiac output and organ blood flow, *Circ. Res.*, 21, 163, 1967.
32. Kaufman, L., Price, D. C., Holliday, M. A., Payne, B., Camp, D. C., Nelson, J. A., and Deconninck, F., Fluorescent excitation analysis in medicine, *J. Radioanal. Chem.*, 43, 321, 1978.
33. Payne, B., Personal communication.
34. Kaufman, L., Shames, D. M., Greenspan, R. H., Powell, M. R., and Perez-Mendez, V., Cardiac output determination by fluorescence excitation in the dog, *Invest. Radiol.*, 7, 365, 1972.
35. Hoffer, P. B., Polcyn, R. E., Moody, R., Lowe, H. J., and Gottschalk, A., Fluorescence detection: application to the study of cerebral blood flow, *J. Nucl. Med.*, 10, 651, 1969.
36. Moody, R. A., Hoffer, P. B., Polcyn, R. E., Lowe, H. J., Gottschalk, A., and Dorben, G. D., K-shell fluorescence for the study of regional cerebral blood flow, *J. Neurosurg.*, 35, 181, 1971.
37. Ter-Pogossian, M. M., Phelps, M. E., Grubb, R. L., Jr., and Gado, M., Measure of regional cerebral blood volume in vivo by means of stimulated X-ray fluorescent x-radiation, and factors affecting this parametèr, *Eur. Neurol.*, 6, 218, 1971/1972.
38. Grubb, R. L., Jr., Phelps, M. E., and Ter-Pogossian, M. M., Regional cerebral blood volume in humans, *Arch. Neurol.*, 28, 38, 1973.
39. Grubb, R. L., Jr., Phelps, M. E., Raichle, M. E., and Ter-Pogossian, M. M., The effects of arterial blood pressure on the regional cerebral blood volume by X-ray fluorescence, *Stroke*, 4, 390, 1973.
40. Phelps, M. E., Grubb, R. L., Jr., and Ter-Pogossian, M. M., In vivo regional cerebral blood volume by X-ray fluorescence: validation of method, *J. Appl. Physiol.*, 35, 741, 1973.
41. Nadel, J. A., Powdered tantalum, a new contrast medium for roentgenographic examination of human airways, *N. Engl. J. Med.*, 283, 281, 1970.
42. Gamsu, G. and Nadel, J. A., New technique for roentgenographic study of airways and lungs using powdered tantalum, *Cancer*, 30, 1353, 1972.
43. Kaufman, L. and Gamsu, G., Fluorescent excitation in the measurement of clearance of heavy metals from the lungs, *IEEE Trans. Nucl. Sci.*, NS-21, 721, 1974.
44. Hermann, G. L. and Kiker, W. E., Remote measurement of fluid volume by X-ray fluorescence, *Med. Phys.*, 2, 20, 1975.
45. Kaufman, L., Nelson, J., Price, D., Shames, D., and Wilson, C. J., Some applications of Si(Li) detectors to clinical problems, *IEEE Trans. Nucl. Sci.*, NS-20, 402, 1973.

46. Aird, R. B. and Zealer, D., Pneumoencephalographic study of cerebrospinal fluid absorptive block mechanisms, *Arch. Neurol. Psychiatry*, 66, 199, 1951.
47. Potts, D. G., Deck, M. D. F., and Deonarine, V., Measurement of the rate of CSF formation in the lateral ventricles of the dog, *Radiology*, 98, 605, 1971.

Chapter 7

STUDIES OF THE KINETICS OF X-RAY CONTRAST AGENTS USING FLUORESCENT EXCITATION ANALYSIS

James A. Nelson

TABLE OF CONTENTS

I. Introduction 129
A. Historic Considerations 129
B. Traditional Methods of Contrast Evaluation 130

II. Fluorescent Excitation Analysis 130
A. Background 130
B. Impact of FEA on the Study of Contrast Media 130
C. Applications of FEA to the Study of Contrast Media 131

III. Summary 133

References 133

I. INTRODUCTION

A. Historic Considerations

Radio-opaque contrast media have existed for nearly as long as man has known about X-rays.[1] Ever since the first contrast studies, there has been interest in the dynamic aspects of contrast media. Precise knowledge of contrast kinetics is essential in the proper design and clinical use of contrast agents and is important in the evaluation of normal and abnormal physiologic function and the clinical pharmacology of contrast media. Despite the importance of understanding the dynamics of contrast agents, much of our knowledge is incomplete and superficial. As Knoefel[2] pointed out in a review of contrast pharmacology in 1965: "Documentation of these agents is often far from what would be desired. It seems that the development of a new agent often consists of (a) preparation of an organic compound with high iodide content, (b) demonstration that the acute lethal toxicity of the compound in animals is low, and (c) introduction of the compound into clinical trial with the subsequent appearance of numerous papers of the testimonial type, without any comparison of the new agents with existing ones." In the 12 years since this statement, serious efforts have been made to rationally design agents with low toxicity,[3,4] to compare new and existing contrast agents,[5,6] and to insure that clinical trials are performed in a safe, well-designed fashion both to protect patients and to generate sound pharmacologic data.[7]

Numerous studies have analyzed the pharmacokinetics of contrast agents in order to better understand basic physiologic mechanisms underlying the clinical usefulness of these drugs.[8-11] The development of computerized tomography, with its high sensitivity to small contrast differences,[12] may lead to an entirely new generation of contrast agents based on more physiologic, less-toxic chemical compounds. Contrast agents other than "organic compounds with high iodide content"[2] may soon be available. Precise kinetic studies will assure a better understanding and more rational use of such agents.

B. Traditional Methods of Contrast Evaluation

Research on contrast media has received renewed attention through the emphasis placed on basic studies by E. Lasser[14,15] and H. Fisher,[13] followed more recently by Amberg,[16] Benness,[17] Berk,[18] and their colleagues. One problem that arose during this fruitful period, however, was the fact that many investigations became traditional in design. Blood samples were obtained at 15-min intervals with bile and urine being collected at 30-min intervals.[16,18] Various loading doses and infusion regimens were developed in order to establish standardized animal kinetic models. Such traditional experiments frequently arise out of such considerations as the ease of the drug assay, the size of the sample required, the animal model being studied, and the results in the related literature; but as Dietschy[19] points out regarding kinetic experiments, "Not uncommonly it appears that the time interval over which the assay was made, the initial concentration of substrate utilized, etc., were chosen arbitrarily." Since visualization of vascular structures, genitourinary structures, or the biliary tract has customarily utilized iodinated materials, chemical methods for iodine determination and radioiodine techniques have been popular.[20,21] The complexity of the studies and the labor-intensive nature of the analytical techniques forced investigators to minimize the number of samples for a given experiment. Thus, a major potential role existed for the new technology of fluorescent excitation analysis (FEA) because it can perform rapid, accurate, nondestructive quantitation of radiographic contrast media in unprepared biological samples.[24]

II. FLUORESCENT EXCITATION ANALYSIS

A. Background

In the early 1950s, Edholm and Jacobsen[25] conceived the idea of quantitating iodine in vivo using monochromatic gamma transmission sources with two different energies. Subsequent work by Sorenson[26] and Mistretta[27] have further pursued this use of the physical properties of dual KVP absorption studies of contrast media. However, with semiconductor detectors, FEA can improve the sensitivity for iodine quantition by a factor of ten as compared to dual KVP techniques.

B. Impact of FEA on the Study of Contrast Media

The use of FEA in pharmacokinetic studies of radiographic contrast agents has several advantages. FEA allows rapid, accurate, reproducible, nondestructive determination of the concentration of several heavy elements in unprepared in vitro biological samples. Experiments can be analyzed rapidly, increasing the flexibility for modifying experimental design on the basis of up-to-date data.[28] This rapid availability of completely analyzed sample results has permitted more detailed pharmacokinetic analysis of radiopaque contrast agents.[11,28-30] While initial applications followed traditional models, the flexibility allowed by FEA quickly led to more sphisticated experimental designs. In general, two major applicatios of FEA were pursued: (1) modeling of drug kinetic studies based on rapid blood sampling, including continual modification of kinetic experiments made possble by the rapid analysis of preliminary experiments in animals that could act as their own controls;[29,30] and (2) in vivo experiments in which tissue concentrations of appropriate contrast media could be observed and correlated with a physiologic event, or incorporated into a kinetic model.[31] In the following section, the results of in vitro applications of FEA analysis will be discussed.

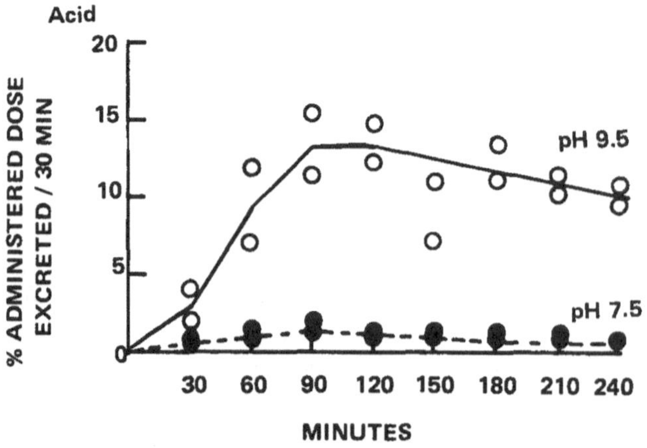

FIGURE 1. Cumulative biliary excretion of iopanoate administered into the duodenum of anesthetized dogs at neutral and alcaline pH. (From Nelson, J. A., Moss, A. A., Benet, L. Z., Goldberg, H. I., and Amberg, J. R., *Invest. Radiol.*, 8, 1, 1973. With permission.

C. Applications of FEA to the Study of Contrast Media

Initial studies of biliary contrast media using FEA were initiated by our group to examine some standing pharmacologic questions. FEA was useful in evaluating the effect of various dose administration schemes as they related to blood and bile concentrations of iodipamide.[32] Similarly, studies of gastrointestinal absorption of iopanoic acid and sodium iopanoate were performed (Figure 1).[33] A rational choice of drug levels and sampling times for optimum evaluation of drug kinetics was now achievable. This led investigators with a more formal background in biopharmaceutical kinetics to become interested in physiologic analysis of contrast media kinetics.[11,28,36]

By observing data on the effects of phenobarbital on iopanoate and iodipamide liver excretion (Figure 2), the saturable nature of iopanoate at clinical dosages was noted.[34] Experiments were designed to test whether multiple constant iopanoate infusions in bile fistula dogs could give quantitative values for a transport maximum. In these experiments, the number of samples that could be used was limited more by the blood volume of the experimental animal than by any other considerations. Frequent sampling of urine and bile was also possible. By obtaining blood samples at frequent intervals, very accurate whole blood concentration curves could be constructed in fine detail for kinetic analysis.[28] Figure 3 demonstrates the type of blood data acquired in these experiments. Each experiment generated 150 to 200 samples, which were typically analyzed by FEA in 1 to 2 days.

The kinetic models derived from this series of experiments have proven beneficial in that they allow accurate and reproducible determination of the hepatic transport maximum of a variety of hepatic contrast agents in healthy animals who can act as their own controls and undergo multiple experiments under varying conditions. Similarly, the quantitative effect of phenobarbital induction has been accurately characterized by this technique,[35] as has the level of endogenous effective bile salt concentration and the effect of cholecystokinin gallbladder contraction.[30] Saturation kinetic models were used to quantitatively compare two biliary contrast agents in the laboratory,[36] expanding on earlier work by Nelson and Staubus.[37] Both efforts indicate methods by which current and future contrast media can be objectively compared prior to clinical use.

An initial approach to iodipamide kinetic modeling has been published,[11] followed by more complex nonlinear kinetic analysis of iodipamide blood curves using the

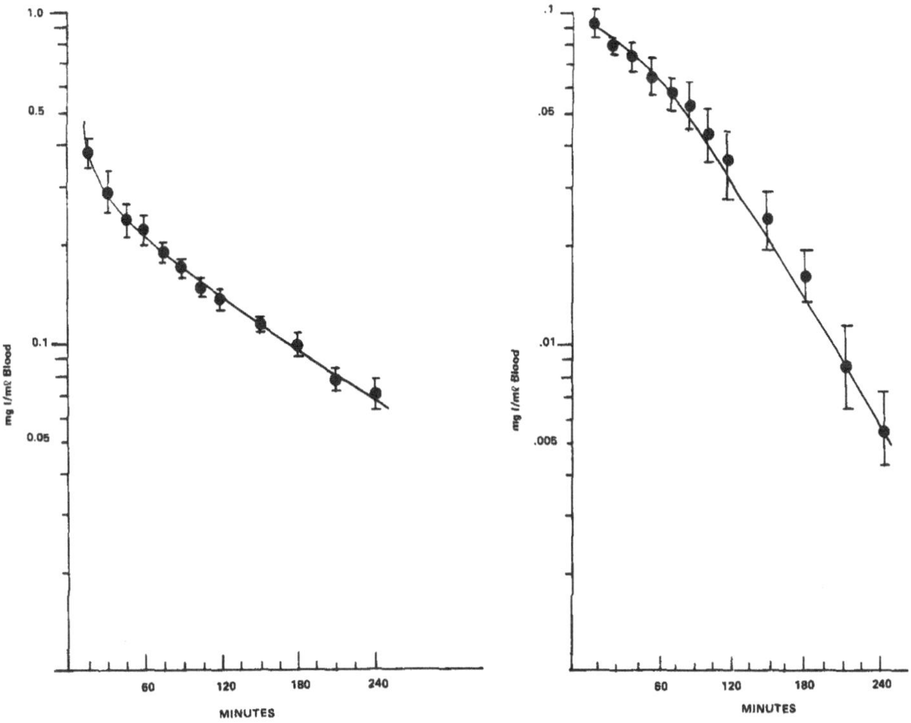

FIGURE 2. Whole blood iodine curve following an i.v. bolus injection of 50 mg/Kg iodipamide. (A) Whole blood iodine curve following an i.v. bolus injection of 50 mg/Kg sodium iopanoate. (B) Studies were performed on unanesthetized bile fistula dogs. (From Nelson, J. A., Pepper, H. W., Goldberg, H. I., Moss, A. A., and Amberg, J. R., *Invest. Radiol.*, 8, 126, 1973. With permission.

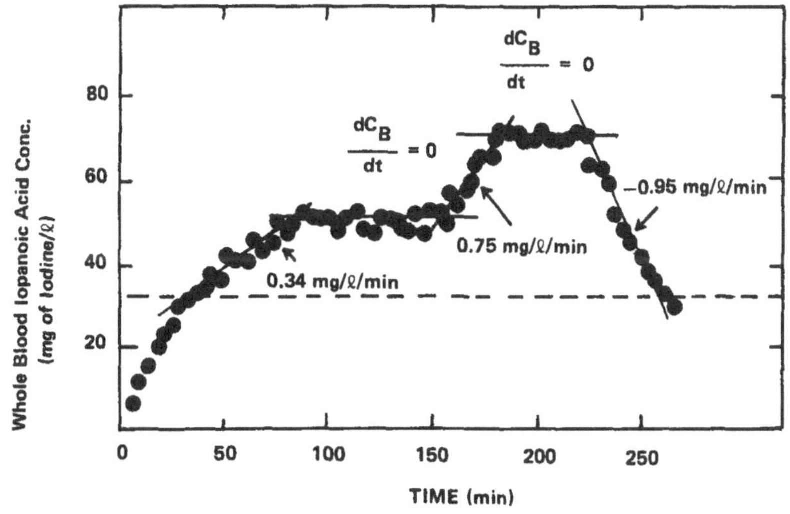

FIGURE 3. Whole blood iodine concentrations in 69 samples. The linearity of iodine concentration with infusion rate is easily demonstrated with frequent sampling. (From Nelson, J. A., Staubus, A. E., Riegelman, S., *Invest. Radiol.*, 10, 371, 1975. With permission.)

SAAM program.[38] This work clearly demonstrates the importance of analyzing unbound and total contrast media in the blood when a wide range of blood concentrations are used. Thus, some blood sample manipulation may become important even with the present simple FEA techniques.

As valuable as in vitro FEA may be for kinetic studies, the capacity of FEA to quantitate contrast material in vivo may ultimately prove to be just as important. Work by Kaufman and his colleagues[39] has demonstrated that an absorption correction technique is applicable for FEA quantitation of iodine in vivo. Subsequent work by Koehler and his colleagues[31] has demonstrated that in vivo iodine determinations allow analysis of accurate liver concentrations in living animals.

FEA also has applications in the analysis of body distribution as well as excretion kinetics of both established and new radiographic contrast media for CT scanning. For instance, Shanser[40] has recently used FEA to study transport of iodinated contrast media into benign renal cysts during pyelographic and arteriographic studies, demonstrating that contrast concentrations in cyst fluid never reach high enough levels to influence cyst density on CT body scanning.

III. SUMMARY

In vitro FEA techniques have made a major impact on the experimental design and analysis of experiments to evaluate radiographic contrast media. Quantitative analysis of the concentrations of contrast media in unprepared or minimally prepared biological samples by FEA is a competitive, nonlabor-intensive, analytical technique which has found ready and widespread application.

The usefulness of quantitative FEA technology is attested to by the number of research centers now using this technology for quantitation of iodinated contrast media and for tracing other heavy elements in both experimental and clinical settings.[41-43] As more investigators become familiar with the technique, its uses will expand. Such work will greatly enhance our basic understanding of the kinetics of various contrast media, will improve utilization of existing contrast media, and will aid in the rational development of new contrast agents.

REFERENCES

1. **Killen, D. A.**, Angiographic contrast media: a historical résumé, *Surgery*, 73, 333, 1973.
2. **Knoefel, P. K.**, Radiopaque diagnostic agents, *Ann. Rev. Pharmacol.*, 5, 321, 1965.
3. **Sovak, M., Nahlovshy, B., Lang, J., and Lasser, E. C.**, Preliminary evaluation of di-ioxo-triglucosyl benzene, *Radiology*, 117, 717, 1975.
4. **Fischer, H. W.**, Improvement in radiographic contrast media through the development of colloidal or particulate media: an analysis, *J. Theor. Biol.*, 67, 653, 1977.
5. **Meyers, P. H., Nice, C. M., Mouton, R. A., Caldwell, T., Meckstroth, G. R., Elder, S. T., and Moser, P. J.**, Comparative evaluation of a new oral cholecystographic agent U-12,031 with telepaque, *Am. J. Roentgenol.*, 94, 491, 1965.
6. **Stanley, R. J., Melson, G. L., Cubillo, E., and Hesker, A. E.**, A comparison of three cholecystographic agents, *Radiology*, 112, 513, 1974.
7. **Russell, J. G. and Frederick, P. R.**, Clinical comparison of tyropanoate sodium, ipodate sodium, and iopanoic acid, *Radiology*, 112, 519, 1974.
8. **Turco, G. L., Ghemi, F., Molino, G., and Segre, G.**, The kinetics of I^{131} rose bengal in normal and cirrhotic subjects studied by compartmental analysis and a digital computer, *J. Lab. Clin. Med.*, 67, 983, 1966.
9. **Dunn, C. R. and Berk, R. N.**, The pharmacokinetics of telepaque metabolism: the relation of blood concentration and bile flow to the rate of hepatic excretion, *Am. J. Roentgenol.*, 114, 758, 1972.

10. Dure-Smith, P., Simenhoff, M., Brodshy, S., and Zimskind, P. D., Opacification of the urinary tract during excretory urography: concentration vs. amount of contrast medium, *Invest. Radiol.*, 7, 407, 1972.
11. Shames, D. M. and Moss, A. A., Iodipamide Kinetics in the dog: a multicompartmental analysis, *Invest. Radiol.*, 9, 141, 1974.
12. Drayer, B. P., Rosenbaum, A. E., Kennerdell, J. S., Robinson, A. G., Bank, W. O., and Deeb, Z. L., Computer tomographic diagnosis of suprasellar masses by intrathecal enhancement, *Radiology*, 123, 339, 1977.
13. Fischer, H. W., Excretion of iodipamide — relation of bile and urine outputs to dose, *Radiology*, 84, 483, 1965.
14. Lasser, E. C., Pharmacodynamics of biliary contrast media, *Radiol. Clin. North Am.*, 4, 511, 1966.
15. Lasser, E. C., Farr, R. S., Fujimagari, T., Tripp, W. N., The significance of protein binding of contrast media in roentgen diagnosis, *Am. J. Roentgenol.*, 87, 338, 1962.
16. Moss, A. A., Amberg, J. R., and Jones, R. S., Relationship of bile salts and bile flow to biliary excretion of iopanoic acid, *Invest. Radiol.*, 7, 11, 1972.
17. Benness, G. T., Urographic contrast agents — comparison of sodium and methylglucamine salts of iothalamate monomer and dimer, *Clin. Radiol.*, 24, 445, 1973.
18. Dunn, C. R. and Berk, R. N., The pharmacokinetics of telepaque metabolism: the relation of blood concentration and bile flow to the rate of hepatic excretion, *Am. J. Roentgenol. Radium. Ther. Nucl. Med.*, 114, 758, 1972.
19. Dietschy, J. M., Difficulties in determining valid rate constants for transport and metabolic processes, *Gastroenterology*, 58, 863, 1970.
20. Zak, B. B.and Boyle, A. J., *J. Am. Pharm. Assoc. Sci. Ed.*, 41, 260, 1952.
21. Goldberger L. E., Berk, R. N., Lang, J. H., and Loeb, P. M., Biopharmaceutical factors influencing the intestinal absorption of iopanoic acid, *Invest. Radiol.*, 9, 16, 1974.
22. Medizihradsky, R., Dahlstron, P. J., Hunter, T. B., Thornbury, J. R., and Silver, T. M., A simple spectrophotometric procedure for the determination of various contrast media in plasma, *Invest. Radiol.*, 10, 532, 1975.
23. Fleck, A., Caine, S., Lucas, J., and Duncan, J. G., Determination of plasma levels of biliary contrast media, *Clin. Radiol.*, 25, 511, 1974.
24. Moss, A. A., Kaufman, L., and Nelson, J. A., Fluorescent excitation analysis: new method of in vitro iodine determination, *Invest. Radiol.*, 7, 335, 1972.
25. Edholm, P. and Jacobsen, B., Quantitative determination of iodine in vivo, *Acta Radiol.*, 52, 337, 1959.
26. Mazess, R. B., Cameron, J. R., and Sorneson, J. A., Determining body composition by radiation absortion spectrometry, *Nature (London)*, 228, 771, 1970.
27. Mistretta, C. A., Ort, M. G., Kelcz, F., Cameron, J. R., Siedband, M. P., and Crummy, A. B., Absorption edge fluoroscopy using quasi-monoenergetic x-ray beams, *Invest. Radiol.*, 8, 402, 1973.
28. Nelson, J. A., Staubus, A. E., and Riegelman, S., Saturation kinetics of iopanoate in the dog, *Invest. Radiol.*, 10, 371, 1975.
29. Kaufman, L., Deconinck, F., Price, D. C., Guesry, P., Wilson, C. J., Hruska, B., Swann, S. J., Camp, D. C., Voegele, L., Friesen, R. D., and Nelson, J. A., An automated fluorescent excitation analysis system for medical applications, *Invest. Radiol.*, 11, 210, 1976.
30. Herzog, R. J. and Nelson, J. A., The role of cholecystokinin in radiographic opacification of the gallbladder, *Invest. Radiol.*, 11, 440, 1976.
31. Koehler, R. E., Kaufman, L., Brito, A., and Nelson, J. A., In vivo measurement of hepatic iodine concentration using fluorescent excitation analysis, *Invest. Radiol.*, 11, 134, 1976.
32. Moss, A. A., Nelson, J., and Amberg, J., Intravenous cholangiography an experimental evaluation of several currently proposed methods, *Am. J. Roentgenol.*, 117, 406, 1973.
33. Nelson, J. A., Moss, A. A., Goldberg, H. I., Benet, L. Z., and Amberg, J., Gastrointestinal absorption of iopanoic acid, *Invest. Radiol.*, 8, 1, 1973.
34. Nelson, J. A., Pepper, H. W., Goldberg, H. I., Moss, A. A., and Amberg, J. R., Effect of phenobarbital on iodipamide and iopanoate bile excretion, *Invest. Radiol.*, 8, 126, 1973.
35. Herzog, R. J., Nelson, J. A., and Staubus, A. E., Saturation kinetics of iopanoate in dogs with an intact enterohepatic circulation before and after phenobarbital induction, *Invest. Radiol.*, 11, 32, 1976.
36. Staubus, A. E., Berk, R. N., Loeb, P., and Cobo-Frenkel, A., Saturation kinetics of iocetamic acid: evaluation of indirect pharmacokinetic techniques and comparison with iopanoic acid, *Invest. Radiol.*, 13, 85, 1978.
37. Nelson, J. A., Staubus, A. E., and Amberg, J. R., The choleretic action of sodium tyropanoate (bilopaque sodium) in the dog: preliminary observations, *Invest. Radiol.*, 9, 438, 1974.
38. Lin, S. K., Moss, A. A., and Riegelman, S., Saturation kinetics of iodipamide, *Invest. Radiol.*, 12, 175, 1977.

39. **Kaufman, L., Shames, D., and Powell, M.**, An absorption correction technique for in vivo iodine quantitation by fluorescent excitation, *Invest. Radiol.*, 8, 167, 1973.
40. **Shanser, J. D.**, Personal communication and paper presented at 1977 RSNA, Chicago, November, 1977.
41. **Alazraki, N., Verba, J. W., Henry, J. E., Becker, R., Taylor, A., Jr., and Halpern, S. E.**, Noninvasive determination of glomerular filtration using X-ray fluorescence, *Radiology*, 122, 183, 1977.
42. **Katzberg, R. W. and Wood, B. P.**, Diatrizoate absorption from enemas in newborn puppies, *Radiology*, 123, 333, 1977.
43. **Thompson, W. M., Jones, R. S., Lang, J. H., and Amberg, J. R.**, Effect of iopanoate on the biliary and urinary excretion of iodipanide, *Invest. Radiol.*, 13, 79, 1978.

Section IV
Thyroid Studies

Chapter 8

CURRENT DEVELOPMENTS IN INSTRUMENTATION FOR FLUORESCENT SCANNING OF THE THYROID

David W. Palmer, Leon Kaufman, and Frank Deconinck

TABLE OF CONTENTS

I. Introduction 139

II. Formulation for Characterization of Fluorescent Scanners 139

III. Results 142

IV. Discussion 145

V. Conclusions 147

References 147

I. INTRODUCTION

Because of its small mass and superficial location, as well as for its high iodine content, the thyroid gland is the only organ for which imaging by fluorescent techniques has found a clinically useful niche. Scans are produced with low doses of radiation, and exposure is limited to the area of the neck, so that fluorescent thyroid scanning is safe for use in children, and in pregnant and lactating women. More important for its general applicability, the distribution of endogenous iodine in disease is not necessarily the same as the distribution of uptake of radioiodine. Thus, new diagnostic information can be obtained.

Fluorescent thyroid scanning was originally introduced as an imaging technique,[1] but was later extended to quantitate intrathyroidal iodine content.[2,3] In essence, a beam of energetic photons from an Am-241 source[1,2] of approximately 10 Ci activity or from an X-ray tube[4] is focused to a fine point within the gland, and an uncollimated or roughly collimated Si(Li)[1,2,4] or HPGe[5] detector of approximately 500 mm^2 in area is used to separate the resulting characteristic radiation from the Compton background (Figure 1). The full assembly is mounted on a rectilinear scanner which travels in a raster pattern to form an image. Iodine quantitation is obtained by monitoring a background window and subtracting the counts in it, NB, from the counts in the Kα window NK (Figure 1), either using a computer[2] or a low-cost system consisting of two single-channel analyzers and an up-down scaler[3] (Figure 2).

II. FORMULATION FOR CHARACTERIZATION OF FLUORESCENT SCANNERS

Image quality is characterized by three important parameters: Spatial resolution (R) (FWHM in mm); sensitivity (S), in counts/sec/ppm I from the full sensitive volume;

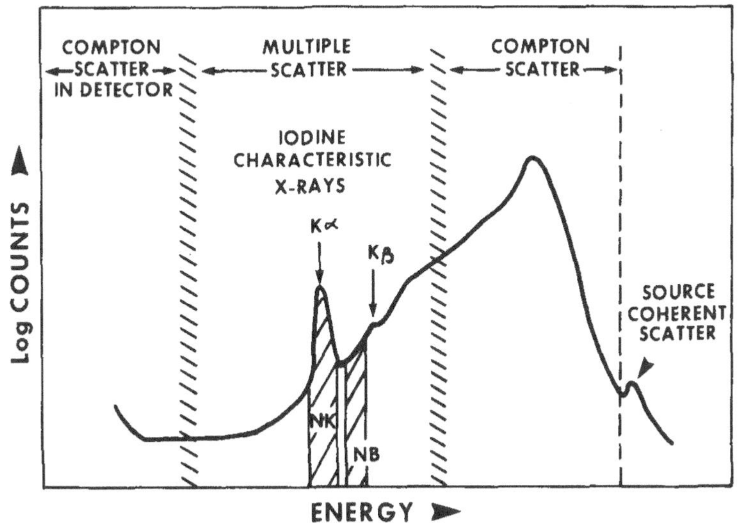

FIGURE 1. Typical spectrum obtained in a Si(Li)-based scanner using an Am-241 source for excitation. The Kβ peak is usually not prominent enough to allow for absorption correction based on the ratio of Kα/Kβ photons.

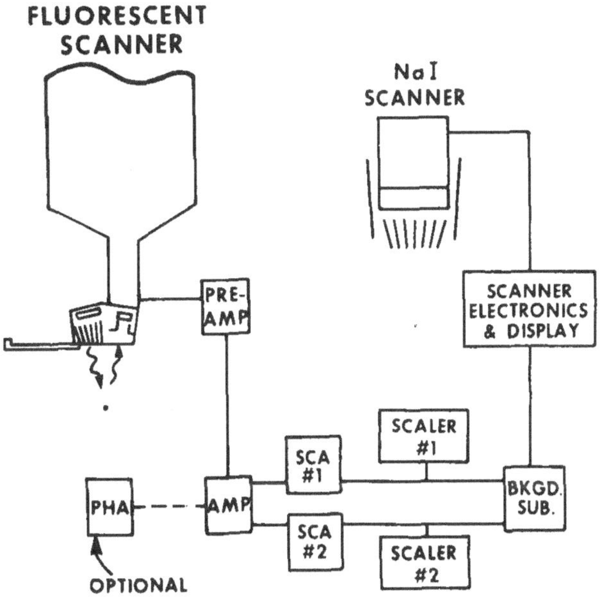

FIGURE 2. Schematic representation of a fluorescent thyroid scanner and a possible electronic configuration for imaging and quantitation.

and background (B), the equivalent ppm I from a blank sample. The background can be obtained from

$$B = \frac{\text{counts/min from a water sample}}{\text{sensitivity in counts/min/ppm I}} \qquad (1)$$

Other important parameters include the spatial symmetry of response of the scanner, the depth of focus, and the radiation exposure rate, which is comparable for the scanners being evaluated. The imaging performance of the scanner can be analyzed in terms of the number of resolved elements per unit area, N, using Rose's formulation:[6]

$$N = \frac{TC^2}{k^2} \qquad (2)$$

where T is the total number of counts per unit area, C is the image contrast of the lesion of interest, defined between 0 and 1, and k is the confidence factor, which for k = 2 yields a 97.7% confidence that the count variation of an element with respect to its surroundings is not a false alarm.

The image contrast, C_o, that would be obtained with an instrument of perfect spatial resolution and zero background, will be decreased through two effects: (1) less than perfect spatial resolution, and (2) background counts in NK. For a cylindrical lesion, the degradation of contrast due to spatial resolution is expressed by $\exp\text{-}(R/D)^2$, where R is the FWHM of a Gaussian point spread function with which we approximate the actual point spread function of the scanner, and D is the lesion diameter. For a cold lesion, the loss of contrast due to background is given by $(1 + B/Wa)^{-1}$, where W and B are the iodine concentration of the tissue around the lesion and the background of the system, both in ppmI, and a is a parameter that accounts for the decreased contribution to the net iodine counts due to absorption effects. We can then write

$$C = C_0 \frac{\exp - (R/D)^2}{(1 + \frac{B}{Wa})} \qquad (3)$$

T in the vicinity of the lesion can be written as

$$T = S\,(Wa + B)\,t \qquad (4)$$

where t is the scan time over the area of interest.

Equation 2 can then be written as

$$N = \frac{S(Wa + B)\,t\,C_0{}^2 \exp - 2\,(R/D)^2}{k^2(1 + \frac{B}{Wa})^2} \qquad (5)$$

where N represents the number of resolved elements over the area covered in the scan time t.

Under identical scan conditions, we can define an arbitrary comparison parameter N′, which contains only the terms that are dependent on the scanner.

$$N' = \frac{S \exp - 2\,(R/D)^2}{(Wa + B)} \qquad (6)$$

For the purpose of comparing different scanners we will use equation 6, and the comparison will be effected in the critical area of low intrathyroidal iodine concentrations, that is, in the limit where W→O, with D chosen at 10 mm. Thus

$$N' = \frac{S \exp - 2\,(R/10)^2}{B} \qquad (7)$$

The quantitation accuracy of the scanner can be obtained as follows: The iodine content, W, is given by

$$W = K\,(NK - W_1) \qquad (8)$$

where K is a calibration constant obtained from known thyroid phantoms, and W_1 represents the counts in a background window just above the iodine peak (Figure 1), which we will assume has been adjusted to yield $W_1 = NK$ for the case of no iodine present in the phantom. This region is chosen because it follows the background count rate variations in the $K\alpha$ peak region more closely than any other region of the spectrum for an expected wide range of scan conditions, e.g., scan width and length variations and different neck topographies.

We can write the uncertainty ΔW as

$$\Delta W = K(NK + W_1)^{1/2} \tag{9}$$

Since the scan is effected over an area larger than that occupied by the gland, the background contribution will be larger than that to be expected if the scan were to be limited to the gland itself. We then obtain $NK = fWSt + BSt$ and $W_1 = BSt$, where t is the scan time and f is the fraction of the scan area occupied by the thyroid. We write equation 9 as

$$\Delta W = K[St(fW + 2B)]^{1/2}$$

To compare quantitation accuracy of various scanners, we define a parameter $\Delta W'$ which in the limit of low concentrations is, for identical scan conditions

$$\Delta W' = K\sqrt{SB}$$

and since $K\alpha 1/s$,

$$\Delta W' = \sqrt{\frac{B}{S}} \tag{10}$$

This formulation will provide comparisons that are most valid for thyroid glands with low concentrations of iodine, which is the case where higher accuracy is needed. In addition, the 10 mm diameter lesion used for comparison purposes in imaging is arbitrary, but reasonable within the context of thyroid imaging.

III. RESULTS

Four fluorescent scanners have recently been evaluated, each representing a different configuration. Table 1 shows the distinguishing features of each unit. Further details

TABLE 1

Fea Scanner Features[*]

Scanner	Detector	Source	Source collimator	Ref.
ORTEC Si(Li)	Si(Li)	Am—241	Focused ring	2
ORTEC HPGe	HPGe	Am—241	Focused ring	5
KeVex	Si(Li)	Am—241	Focused planar	1,3
PGT	Si(Li)	X-ray tube	Straight bore	4

* ORTEC Si(Li) Scanner evaluated courtesy of Dr. Joel I. Hamburger, M.D., Northland Thyroid Laboratory, Southfield, Michigan. ORTEC HPGe Scanner evaluated courtesy of Dr. Bruce A. Mazat, M.D. Tripler Army Medical Center, Hawaii. PGT Scanner evaluated courtesy of Dr. Peter Esser, Ph.D., Presbyterian Hospital, New York, New York. KeVex Si(Li) detector evaluated at the University of California, San Francisco, California.

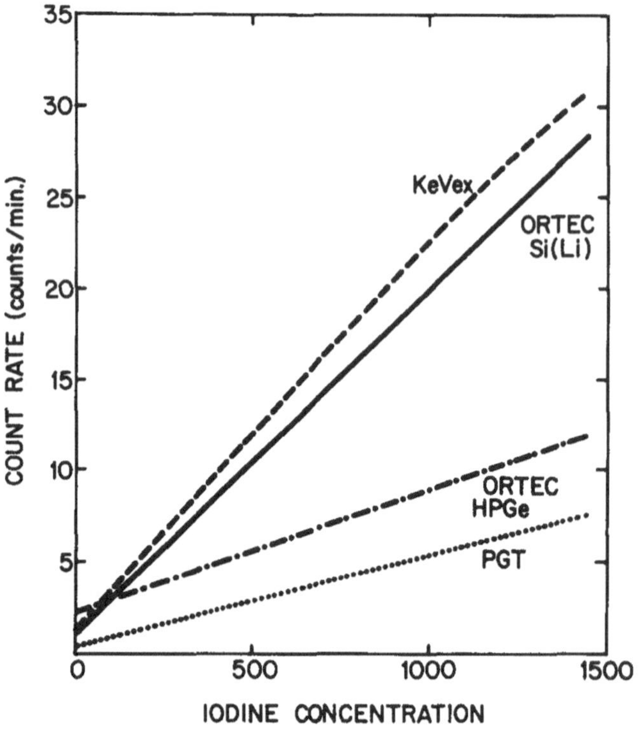

FIGURE 3. Sensitivity of various scanners, measured with iodine sources of 2 cm-diam placed at the point of maximum response.

can be found in references.[1-5] Intrinsic spatial resolution was measured with a line-source of iodine with a concentration of approximately 30% by weight. The line spread function was measured by moving the line-source in the plane of the scan under the fixed scanner, this process being performed at various depths from the collimators. For scanners with an asymmetric excitation source-detector configuration,[1,4] the measurements were also performed along orthogonal directions in the plane of the scan, but no significant differences were found. Sensitivity was measured by observing the counts within the NK window using 20 mℓ vials of approximately 2 cm-diam, filled with iodine solutions ranging from zero to 1,500 ppmI, placed horizontally with the center of the vial at the point of maximum response. NK window widths were set at the full width at one tenth maximum level of the Kα-line, following criteria discussed in Reference 7 (See also Chapter 5). Window width results from the combined effects of detector resolution (approximately 600 to 650 eV at this energy) and the intrinsic separation of the Kα_1-Kα_2 lines of iodine (295 eV). The detector systems were otherwise configured as the manufacturers delivered them. These measurements yielded sensitivity, which is just the slope of the count-rate vs. concentration curve (Figure 3), and the background, calculated from equation 1.

Intrinsic spatial resolution as a function of depth is shown in Figure 4. The focal point was defined as the region of maximum count-rate response from the line source. Note that this does not necessarily coincide with the point of highest resolution. Note also the excellent performance of the Princeton Gamma-Tech (PGT) scanner, and the shallow response and sensitivity curves in the KeVex® scanner when compared to the Oak Ridge Technical Enterprises Corp. (ORTEC) units. The latter degrade rapidly in resolution as one moves away from the focal spot (Figure 5). Since the thyroid has a

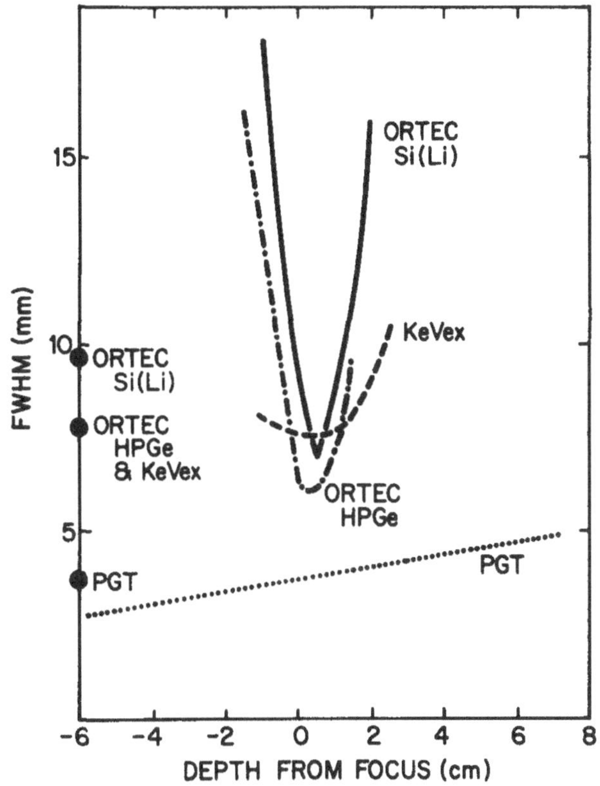

FIGURE 4. Spatial resolution in the plane of the scan as a function of depth. The black dots indicate the spatial resolution averaged over a 2 cm-deep region.

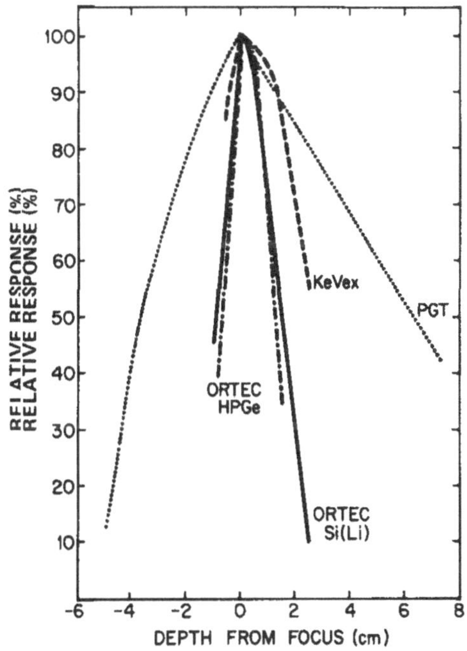

FIGURE 5. The sensitivity of various scanners varies as a function of depth. A shallow response curve is desirable.

TABLE 2
Fea Scanner Performance*

Scanner	Sensitivity cts/ min/ppmI	Background cts/min	Background ppmI	Resolution FWHM, mm	N′	ΔW′
ORTEC Si(Li)	18.8	909	48.3	9.6	10	1
ORTEC HPGe	6.74	2184	324	7.8	1	4.3
KeVex	21.1	1129	53.5	7.7	19	1
PGT	5.13	263	51.2	3.7	12	2

ORTEC Si(Li) Scanner evaluated courtesy of Joel I. Hamburger, M.D., Northland Thyroid Laboratory, Southfield, Michigan. ORTEC HPGe Scanner evaluated courtesy of Dr. Bruce A. Mazat, M.D., Tripler Army Medical Center, Hawaii. PGT Scanner evaluated courtesy of Dr. Peter Esser, Ph.D., Presbyterian Hospital, New York, New York. KeVex Si(Li) detector evaluated at the University of California, San Francisco, California.

finite depth, we used as the line spread function the average value over a 2 cm-deep region. Because a 2 cm-diam source was used to measure sensitivity, these results are automatically averaged for depth response. N′ and ΔW′ were calculated from equations 7 and 10, respectively, and the results were summarized in Table 2.

IV. DISCUSSION

Intrathyroidal iodine quantitation has become a significant diagnostic component in the characterization of thyroid disease. Our data indicate that ORTEC Si(Li) and KeVex detectors are essentially identical for this purpose, with the PGT and ORTEC HPGe systems considerably less adequate. To compensate for the loss of quantitation accuracy associated with gland depth, patients can be separated into two classes, lean and obese, the classification depending upon the amount of tissue overlying their thyroids, and values of K determined for each class. We have performed scans of a series of thyroid phantoms of different iodine content, partially immersed in water and covered with tissue-equivalent absorbers to simulate the surrounding neck. Linear relationships were found (Figure 6) between the total iodine content of the phantoms and the net iodine Kα counts, validating the assumed proportionality of the equation.

In practice, following a thyroid scan, the patient is classified as either lean or obese (based upon palpation of the neck to estimate depth of the thyroid), and his intrathyroidal iodine is derived from either the upper or lower curve in Figure 6, which gives rise to an uncertainty corresponding to approximately half of the spread, or 12%.

N′, a measure of the number of resolved elements in a scan, varies over a factor of 20 between the best unit (KeVex®) and the worst (ORTEC HPGe). N′ also varies as a function of lesion diameter, and Figure 7 shows this parameter for various lesion sizes. Because of its superior spatial resolution, the PGT system provides the best imaging performance for lesions of less than 8 mm in diameter, and the KeVex system has the best performance for larger lesions. It would appear that if the sensitivity of the PGT detector could be tripled, its performance would excell. Limitations may be imposed by the power ratings of the X-ray tube or by the nonmonochromatic nature of the excitation beam, which may increase the Compton-to-fluorescent photon ratio reaching the detector. In the latter case, dead-time effects would not allow for an increase in the intensity of the excitation beam.

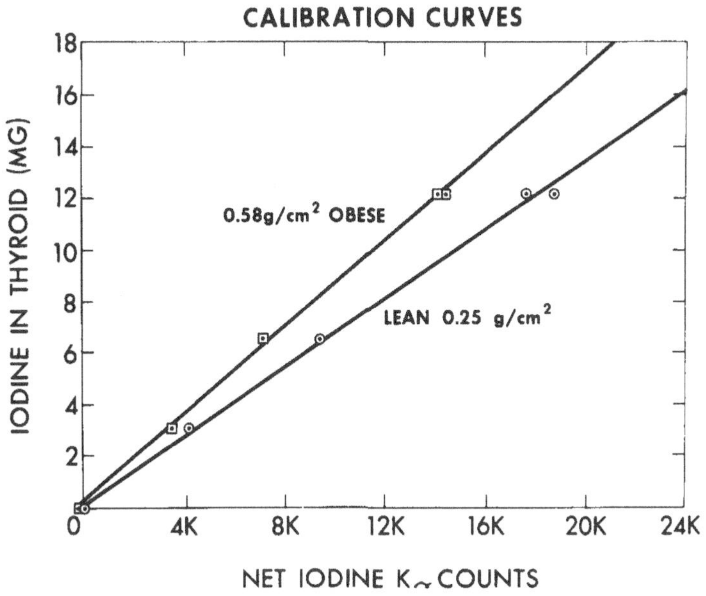

FIGURE 6. Calibration curves obtained with a KeVex thyroid scanner, using a neck phantom and two different skin thicknesses.

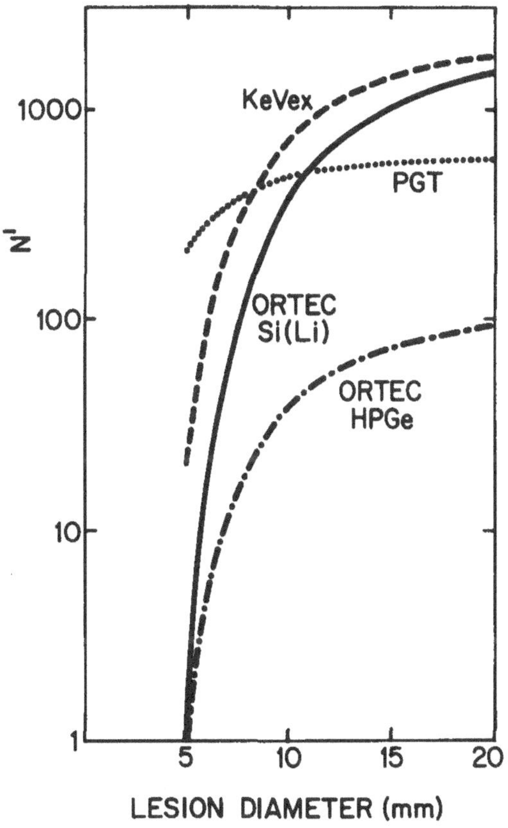

FIGURE 7. N′, a parameter indicative of the number of resolved elements per unit area in a typical scanning situation, is shown for the four units tested.

The performance of the HPGe detector presents itself as an interesting manifestation of the tradeoffs between detection efficiency, sensitivity, and imaging performance. The 5 mm-thick Si(Li) detectors used here are approximately 75% efficient for detection of the iodine $K\alpha$-lines, while the HPGe detectors are 100% efficient. It would then appear that a 33% gain in sensitivity over Si(Li) can be achieved by using HPGe detectors.[8] The problem is, as we have demonstrated earlier, that while the stopping power of Si(Li) detectors for Compton-scattered photons ($\sim$50 keV) is under 40%, it is 100% for HPGe. Since most of the detected events are Compton-scattered photons, dead-time effects from these will be 2.5 times worse in HPGe than in Si(Li). In addition, since background appears to be a fixed fraction of the photons found in the Compton peak, the higher Compton-to-fluorescence detection efficiency of HPGe will raise the background levels in the latter.[9] These previous observations are confirmed by the present results. That dead-time is an important factor is evidenced by the fact that a KeVex scanner, when operated with a 1 μsec rather than 2 μsec shaping time realizes an increase of 50% in net count-rate for iodine, to 31.2 cts/min/ppmI. Because of the poorer energy resolution, which necessitates a wider NK window, the background increases to 69 ppmI. Even so, both the imaging performance and quantitation accuracy increase significantly.

It is thus evident that it is more appropriate to increase effective detection efficiency through the use of optimized electronics and if necessary larger area Si(Li) detectors, say from 25 mm to 29 mm in diameter, than to use HPGe in this application. The loss of energy resolution associated with these steps is not significant for iodine fluorescence. The use of more than one Si(Li) detector is an effective, albeit expensive, means of increasing sensitivity.

V. CONCLUSIONS

The KeVex and ORTEC Si(Li) fluorescent scanners offer equivalent performance for the quantitation of intrathyroidal iodine content. For imaging purposes the KeVex scanner offers better performance for the typical lesions sought with these instruments, while the PGT scanner is to be preferred if imaging of small lesions is of importance and scanning time is not of concern. The ORTEC HPGe scanner has a level of performance well below that of the other systems.

REFERENCES

1. Hoffer, P. B., Bernstein, J., and Gottschalk, A., Fluorescent techniques in thyroid imaging, *Semin. Nucl. Med.*, 1, 379, 1971.
2. Patton, J. A., Brill, A. B., Blanco, G., and Highfill, R., Experiences with semiconductors in imaging and functions studies at Vanderbilt, in Kaufman, L. and Price, D. C. Eds., *Semiconductor Detectors in Medicine*, USAEC. CONF 730321, U.S. Atomic Energy Commission, Office of Information Services, Technical Information Center, Oak Ridge, Tenn., 1973, 253.
3. Palmer, D. W., Deconinck, F., Swann, S. J., Okerlund, M., Kaufman, L., and Hoffer, P. B., Low cost intrathyroidal iodine quantitation with a fluorescent scanner, *Radiology*, 119, 733, 1977.
4. Esser, P. D. and Lister, D. B., A new apparatus for fluorescent scanning: a moving X-ray tube, *J. Nucl. Med.*, 18, 640, 1977.
5. Patton, J. A. and Brill, A. B., Simultaneous emission and fluorescent scanning of the thyroid, *J. Nucl. Med.*, 19, 464, 1978.
6. Rose, A., *Vision: Human and Electronic*, Plenum Press, New York, 1971.

7. **Kaufman, L., Deconinck, F., Price, D. C., Guesry, P., Wilson, C. J., Hruska, B., Swann, S. J., Camp, D. C., Voegele, A. L., Friesen, R. D., and Nelson, J. A.**, An automated fluorescent excitation analysis system for medical applications, *Invest. Radiol.*, 11, 210, 1976.
8. **Martini, M.**, Recent developments in semiconductor radiation detectors, in *Semiconductor Detectors in Medicine*, Kaufman, L. and Price, D. C., Eds., USAEC CONF-730321, U.S. Atomic Energy Commission, Office of Information Services, Technical Information Center, Oak Ridge, Tenn., 1973, 53.
9. **Kaufman, L., Wilson, C. J., Nelson, J. A., and Shames, D. M.**, Techniques for In Vitro and In Vivo Elemental Quantitation by Fluorescent Excitation. Ibid, page 126.

Chapter 9

THE CLINICAL UTILITY OF FLUORESCENT SCANNING OF THE THYROID

Michael D. Okerlund

TABLE OF CONTENTS

I. Introduction 149

II. The Normal Fluorescent Thyroid Scan 150

III. Graves' Disease (Diffuse Toxic Goiter) 151

IV. Nonfunctioning Nodules 151

V. Functioning Thyroid Nodules 151

VI. Hashimoto's Thyroiditis 155

VII. Nodular Goiter in North America 156

VIII. Radiation-Induced Thyroid Disease 156

IX. The Future of Fluorescence Excitation in the Study of Thyroid Physiology 157

References 160

I. INTRODUCTION

In 1964 Heedman and Jacobson[1] first quantitated human stable thyroidal iodine by in vivo X-ray spectrometry. Hoffer[2-4] initially used Dysprosium-159 (Dy-159) and a lithium-drifted silicon crystal in a low temperature vacuum chamber to image the thyroid by FEA. The energy of ^{159}Dy radiation was above the K-shell binding energy of stable iodine and the characteristic X-rays produced in filling the resulting orbital vacancies could be detected externally. Americium sources proved more advantageous in later studies and attachment of the fluorescent apparatus to medical rectilinear scanners produced satisfactory thyroid images and demonstrated the size, shape, and positional relationships of the thyroid lobes. This method produced satisfactory localization of palpable thyroid nodules. A particular advantage of the fluorescent technique is the absence of whole body, gonadal, or bone-marrow radiation and the absence of radiation to the fetus of pregnant women.

More recently, fluorescence excitation has been used to quantitate stable intrathyroidal iodine.[5-8] These studies have indicated slight sex and age differences in stable iodine content, have initiated the study of the relationship of this iodine mass to thyroid physiology, and have produced clinically useful data in various thyroid disorders. Summarized results from thyroid stable iodide quantitation in a large number of patients with common thyroid diseases are shown in Table 1.

II. THE NORMAL FLUORESCENT THYROID SCAN

The types of images obtained from fluorescent thyroid scanning are indistinguishable from those of isotope scanning, since the same electronics and data presentation systems are used. The size, shape, and positional relationships of the thyroid lobes are therefore comparable to isotope scanning, with the right lobe larger in the majority of normal cases, the two lobes of equal size in a smaller number, and the presence of a larger left lobe in a still smaller number. In our system, the ratio of target-to-background count rates is approximately 20 to 1 when the normal thyroid gland is compared to extrathyroidal cervical count rates. The background count rate is due to scatter, the semiconductor detector, and system electronics rather than actually detectable nonthyroidal stable iodine, since the stable iodine content of both nonthyroidal human tissue and blood is extremely low. With regard to iodine quantitation, we have found that the mean value of glandular stable iodine in normal males is approximately 10 mg and the mean female level approximately 9 mg (Table 1). The significance of this apparent difference is unknown, but there are no clinically significant differences between thyroid function as inferred from levels of thyroidal radioiodide uptake, serum thyroxine, serum triiodothyronine, and serum thyrotropin (TSH) between groups of normal males and females. Patton et al.[8] found a mean thyroidal stable iodine of 10.7 ± 4.8 mg in a group of normal persons.

Satisfactory studies can usually be performed in normal individuals taking fully suppressive doses of thyroid medication (2 gr of desiccated thyroid or 0.2 mg of sodium-levo-thyroxine daily) if the period of medication is brief, 4 weeks or less, but in normal persons on long term doses of this magnitude (three months or longer), intrathyroidal stable iodine levels are usually very low and fluorescent thyroid scans generally unsatisfactory. Our initial studies indicate that the mean value of thyroidal stable iodine may decline somewhat with age, but the clinical significance of this finding is unclear at present. In a series of normal patients, there was no significant correlation of glandular stable iodine content with thyroidal isotope uptake, serum thyroxine, serum tri-

TABLE 1

Stable iodine levels detectable in the thyroid glands of consecutive patients with various thyroid disorders

	Number	Range	Mean	S.D.
Normal				
Female	41	4.1—19.0	9.2	4.3
Male	26	3.8—17.5	9.9	4.5
[a]Graves disease	68	6.2—28.0	15.5	7.8
[a]Nodular goiter	53	4.1—20.0	11.7	5.6
[a]Hashimoto's thyroiditis				
Euthyroid	56	ND—18.0[b]	4.8	3.7
Hypothyroid	13	ND—5.2[b]	2.3	2.0
Secondary hypothyroidism	4	6.1—10.6	9.1	2.0

[a] 80—90% Female in each of these groups.
[b] ND = Not detectable

idothyronine, or the serum ratio of the two latter hormones. We have noted no racial differences comparing Caucasian, Negro, and Chinese patients living in North America.

III. GRAVES' DISEASE (DIFFUSE TOXIC GOITER)

In cases of hyperthyroid Graves' disease, the total thyroidal stable iodine content is generally increased in proportion to size of the patient's goiter, implying a concentration of stable iodine which is not greatly different from normal values. Patton and coworkers[8] also described increased glandular iodine in hyperthyroid patients, the latter presumably representing Graves' disease. Euthyroid patients with the ophthalmopathy of Graves' disease behave differently in this respect than patients with hyperthyroidism alone, in that glandular stable iodine levels range from below normal (but detectable) to slightly above the mean normal. The significance of this finding is presently uncertain, but abnormalities of thyroid function, including failure to suppress when treated with exogenous thyroid medication, and flattening of the pituitary response to thyrotropin-releasing-hormone (TRH) are well documented in this group of patients.[9-10]

Patients with surgically treated Graves' disease and euthyroid function usually have normal or low glandular levels of stable iodine, and in patients who are hypothyroid after such surgery, levels are generally low or undetectable. Levels are also usually low or undetectable in patients rendered hypothyroid by therapeutic $NaI,^{131}$ and low or normal in the patients remaining euthyroid after this form of surgery.

Figure 1 shows a patient with hyperthyroidism due to Graves' disease in which the fluorescent scan shows normal stable iodine as a result of previous treatment with Propylthiouracil.

There is some tendency for patients with low levels of glandular iodine to have increases in the ratio of serum triiodothyronine to serum thyroxine in Graves' disease, perhaps indicating some ability of the thyroid gland to conserve iodine by a change in relative secretion of the two hormones.

Figures 2 and 3 show two patients treated, respectively, with surgery and therapeutic radioiodide, both of whom show satisfactory isotope scintigrams, but with glandular iodide depletion shown on the fluorescent study. Both situations are compatible with rapid turnover of a small glandular iodine pool.

IV. NONFUNCTIONING NODULES

Thyroid nodules found in this laboratory to be nonfunctioning by imaging with sodium ^{123}I or Tc-99m pertechnetate have invariably contained no detectable stable iodine by the flourescent technique. Since the resolution obtained by these methods is of the order of 1 cm, it appears in most cases unwarranted to attempt resolution of very small nodules by the fluorescent technique instead of conventional isotope scanning. We have observed no differences between benign and malignant thyroid nodules by this technique, nor in the total iodine content of the glands from which they arise. Hoffer[3,4] also found no stable iodine in both benign and malignant nonfunctioning nodules. The resolution achieved also makes fluorescent thyroid scanning a poor method of searching for functioning thyroid carcinoma since the stable iodine content of this tissue is low in relation to normal tissue, and the lesions tend to be located deep in cervical lymph nodes, or in other poorly accessible locations such as mediastinum, lung, or bone.

V. FUNCTIONING THYROID NODULES

Nodules found to be functioning with radioiodide or pertechnetate contain stable

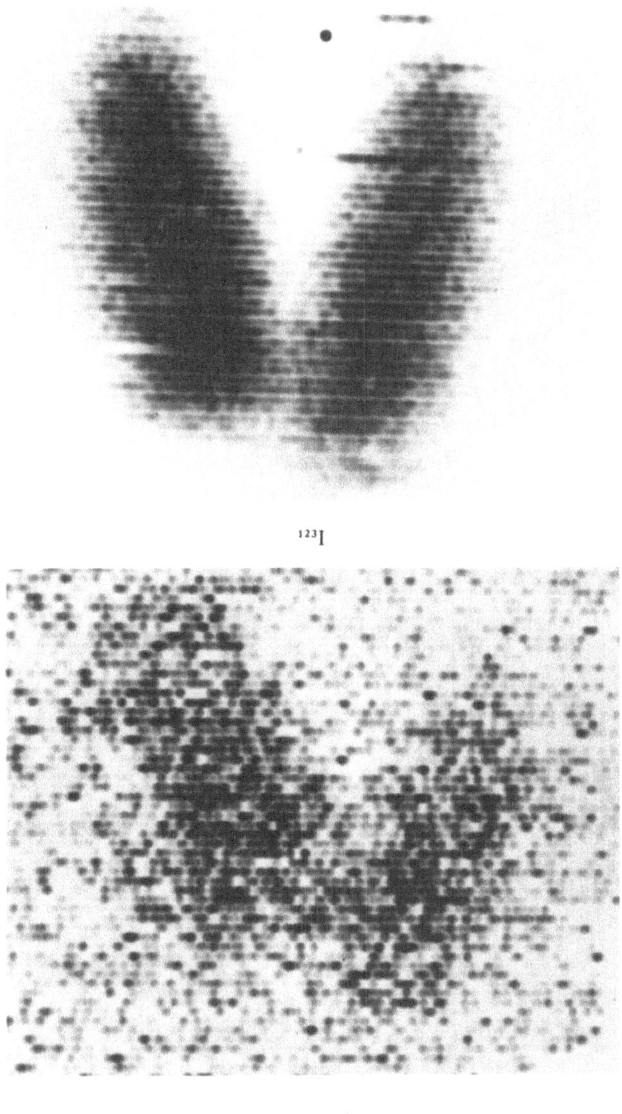

FIGURE 1. Thyroid imaging studies on a 27-year-old female with active Graves' disease despite treatment with propylthiouracil until the day before. She had diffuse thyroid enlargement, a 24-hr thyroidal radioiodide uptake of 100%, and a serum thyroxine of $17 \mu g d\ell$. Rectilinear scintigram (I^{123}) is shown on the left. The fluorescent scan on the right shows a normal stable iodine level of 11 mg, a normal value probably as a result of blockade of iodide organification by antithyroid drug.

iodine in almost all cases, and this content is generally in proportion to the amount of tissue mass present. Large single autonomous thyroid nodules may contain as much as 10 to 15 mg total stable iodine when studied by this technique. In hyperfunctioning nodules with functional suppression of surrounding or contralateral thyroid tissue, the latter tissue cannot usually be located or documented by fluorescent imaging.

Both functioning and nonfunctioning nodules (if greater in largest diameter than approximately 1 cm) can be marked in vivo under the detector for unambiguous localization and correlation with physical location.

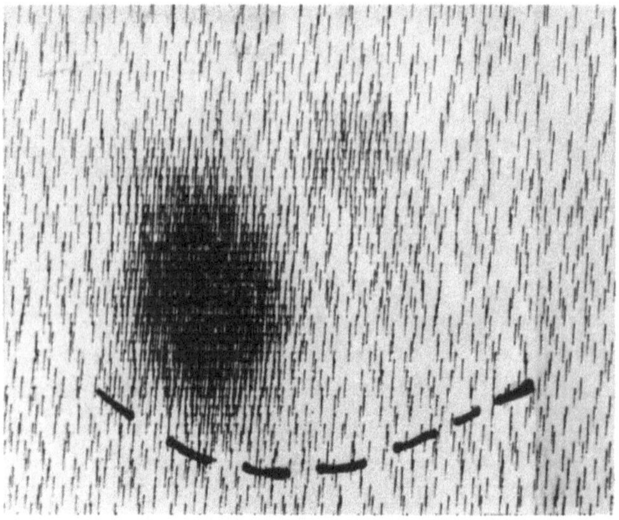

^{123}I

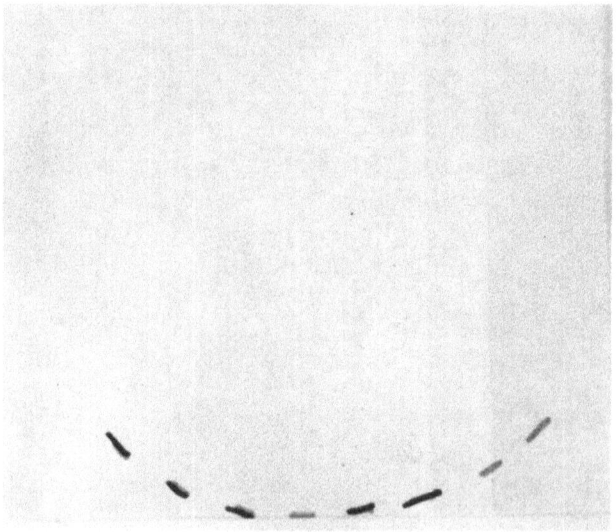

FEA

FIGURE 2. Thyroid imaging studies on a 54-year-old female who had surgery for Graves' disease 24 years previously, and recently developed symptoms suggestive of hypothyroidism. A small amount of thyroid tissue was palpable in the neck. Serum thyroxine was low at 4.7 ug/dℓ, but serum triiodothyronine was normal at 165 ng/dℓ. Thyroidal radioiodide uptake at 24 hr was low normal at 14%. Isotope scan (^{123}I) showed functioning tissue in a residual right lobe and possibly a pyramidal lobe, but fluorescent scan showed less than 1.5 mg stable iodine. Rapid turnover of intrathyroidal iodine was demonstrated by labelling the thyroid with ^{131}I.

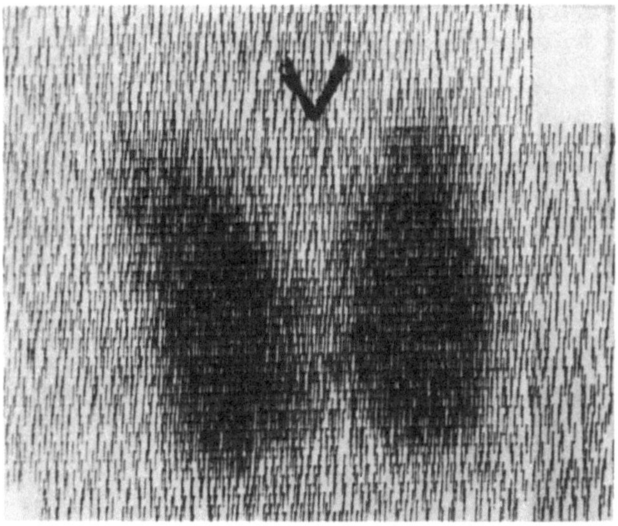

^{123}I

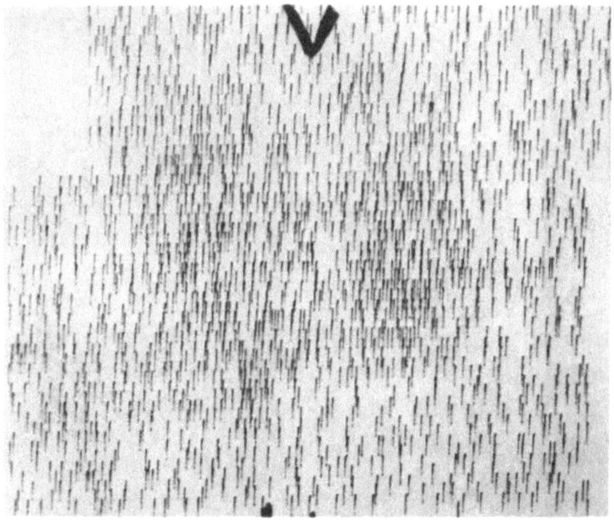

FEA

FIGURE 3. Thyroid imaging studies on a 30-year-old male with persistent Graves' disease despite treatment with therapeutic sodium iodide-131 6 months previously. Serum thyroxine was elevated at 17.8 ug/dℓ and 24-hr thyroidal radioiodide uptake persistently elevated at 46%. Rectilinear scintigram with ^{123}I is shown on the left. The fluorescent scan indicated the presence of only 1.7 mgm stable iodine in the gland, although the level had been greatly elevated (28 mg) before radioiodide therapy.

Figure 4 shows isotope and fluorescent scans from a patient with a large autonomous hyperfunctioning thyroid nodule, showing two very similar images.

VI. HASHIMOTO'S THYROIDITIS

Total thyroidal stable iodine content is decreased in many cases of Hashimoto's thyroiditis and undetectable in some of these. Whether this is due to actual destruction of tissue or kinetic abnormality of iodine metabolism is uncertain at present and may even differ among various types of glands. Values tend to be particularly low in patients with hypothyroidism of this etiology, but may be below normal (although very

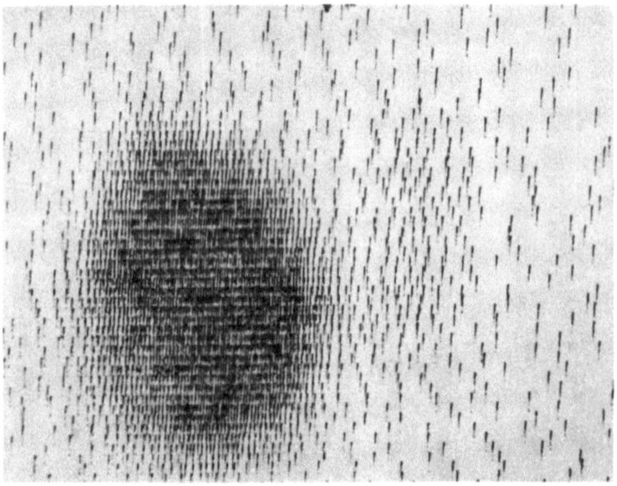

^{123}I

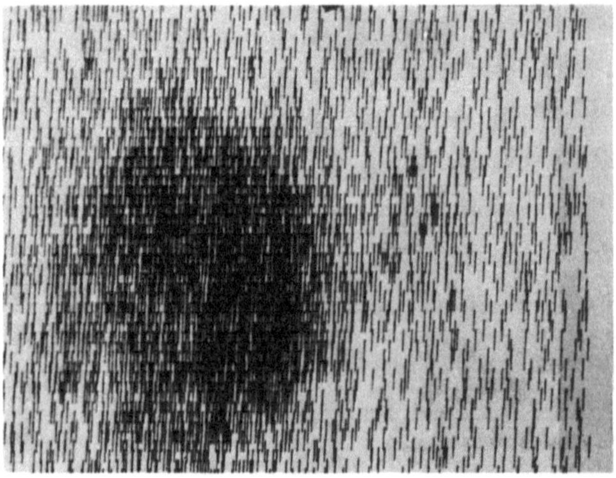

FEA

FIGURE 4. Thyroid imaging studies in a 23-year-old female with hyperthyroidism due to a large autonomous adenoma occupying most of the right lobe. Serum thyroxine was elevated at 21 μg/dℓ and 24-hr thyroidal radioiodide uptake was 79%. Both the isotope (I^{123}) and fluorescent scans show a functioning nodule, but no evidence of the contralateral suppressed lobe, which was palpable clinically.

low on a concentration basis) in patients with goiter without hypothyroidism. We have noted no differences in relation to age, sex, or the presence of a positive family history of the disease in thyroiditis patients. The presence of a negative correlation between total glandular iodine content and the serum ratio of triiodothyronine to thyroxine may indicate some ability to conserve glandular iodine in this disease. However, no relationship has been established between total glandular iodine content and thyroidal radioiodide uptake.

A very low thyroid stable iodine content in a hyperthyroid patient renders secondary (pituitary) hypothyroidism unlikely, as thyroidal stable iodine content in that disorder is usually normal in our experience.

Figure 5 shows a satisfactory isotope scan from a patient with Hashimoto's thyroiditis whose fluorescent scan was unsatisfactory because of iodide depletion. The latter provided diagnostic information. However, Figure 6 shows a normal isotopic and normal fluorescent scan in secondary hypothyroidism. In contrast, Figure 7 illustrates the poor radioiodine uptake and equally poor ^{123}I scan in a patient with subacute thyroiditis whose fluorescent iodine scan was normal.

VII. NODULAR GOITER IN NORTH AMERICA

In almost all cases of this disease, the total glandular iodine content is normal or slightly elevated, although the iodine per gram of tissue must be decreased in the majority of these cases since the thyroids are characteristically enlarged. Levels appear roughly correlated with gland size, however, and appear no different in cases associated with nonfunctioning or functioning nodules of moderate size or less. We have noted no relationship between glandular stable iodine and serum triiodothyronine-to-thyroxine ratio in nodular goiter.

Fluorescent thyroid scanning is generally inferior to the isotopic technique (particularly the use of sodium ^{131}I) for localization of substernal goiters, since the linear attenuation coefficient for the iodine characteristic X-ray is very high and interposed bone renders its consistent detection unlikely. The stable iodine content of large nodular goiters is frequently irregular in distribution, and the ability to identify and localize small nodules is much less satisfactory than higher resolution techniques, such as pinhole gamma camera imaging after intravenous injection of Tc-99m pertechnetate.

VIII. RADIATION-INDUCED THYROID DISEASE

We have previously described a tendency toward low levels of total thyroidal stable iodine in patients who had incidental radiation to the thyroid gland when receiving head or neck radiation for acne, tonsillitis, or thymic enlargement early in life.[11] This group of patients has been of intense interest to thyroidologists because of the high prevalence of thyroid tumors (both benign and malignant) arising many years after radiation in childhood or adolescence.[12-14] In some of these radiation cases, the patients are found to have Hashimoto's thyroiditis, and the glandular iodine depletion may be on that basis. In other cases, this disease appears not to be present and no precise explanation is known. Since pituitary thyrotropin (TSH) is implicated in the development of most thyroid tumors,[15,16] and the iodide depleted thyroid gland is known to be physiologically more sensitive to TSH,[17] any glandular iodide depletion caused by radiation might be clinically significant in the development of tumors in these patients and deserves further study.

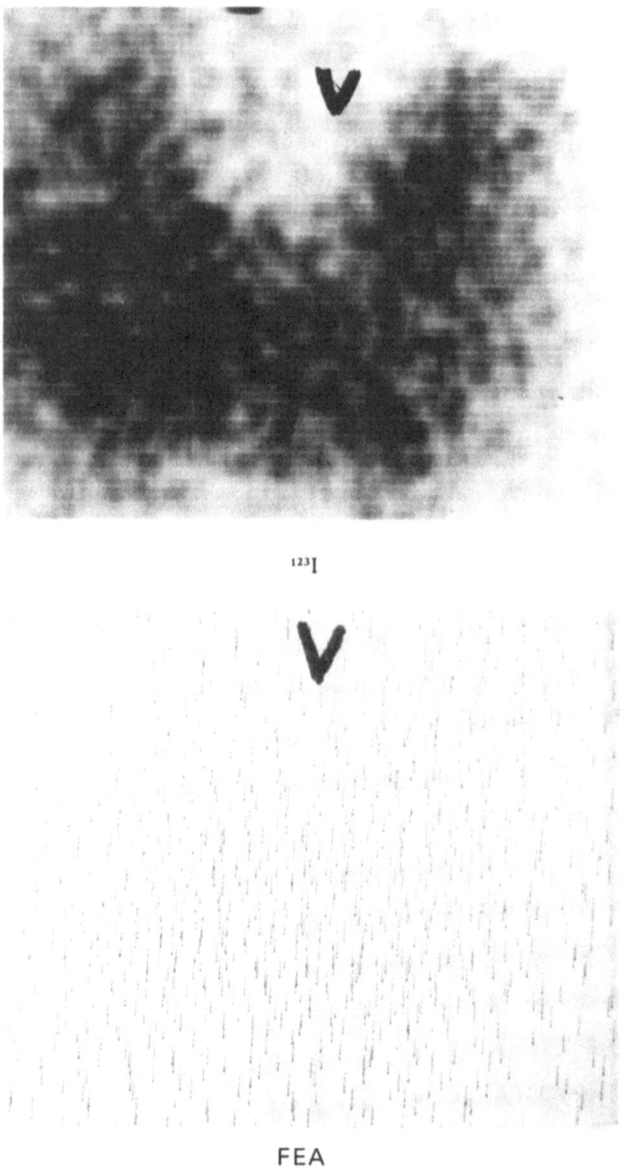

FIGURE 5. Thyroid imaging studies on a 56-year-old female with goitrous hypothyroidism. Serum thyroxine was low at 2.6 μg/dℓ and 24-hr thyroidal radioiodide uptake normal at 28%. Thyroglobulin and microsomal antibodies were both present in high titer in the serum. Rectilinear scintigram (^{123}I) showed glandular enlargement, but fluorescent excitation scan showed no detectable stable iodine. The final diagnosis was Hashimoto's (chronic lymphocytic) thyroiditis.

IX. THE FUTURE OF FLUORESCENCE EXCITATION IN THE STUDY OF THYROID PHYSIOLOGY

Future developments in excitation sources, semiconductor detectors, and electronics may result in improvement in the results now available with fluorescence excitation studies of the thyroid gland. Increases in resolution and sensitivity may also increase the diagnostic yield regarding small thyroid nodules. Even a two-to-four-fold improvement in realizable resolution would produce thyroid images demonstrating thyroid

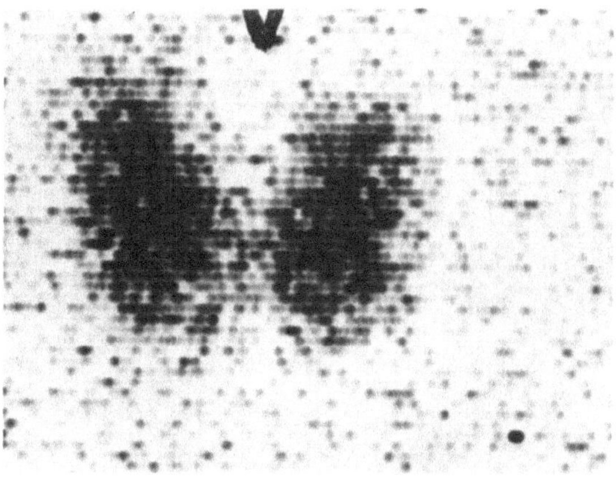

FEA

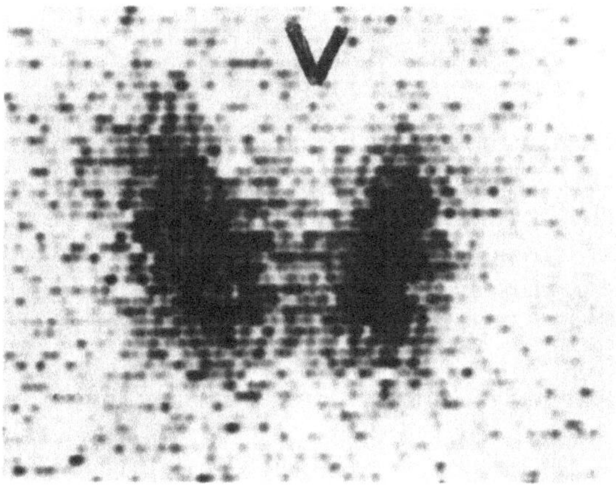

^{123}I

FIGURE 6. Thyroid imaging studies on a 25-year-old female with clinical signs of hypothyroidism, amenorrhea, and a chromophobe adenoma of the pituitary gland. Serum thyroxine was low at 5.1 $\mu g/d\ell$ and serum TSH only 3mU/mℓ despite hypothyroidism; 24-hr thyroidal radioiodide uptake was low normal at 12% and rectilinear scintigram (^{123}I) was normal. The normal fluorescent scan and stable thyroidal iodine levels provided evidence that the hypothyroidism was pituitary in origin.

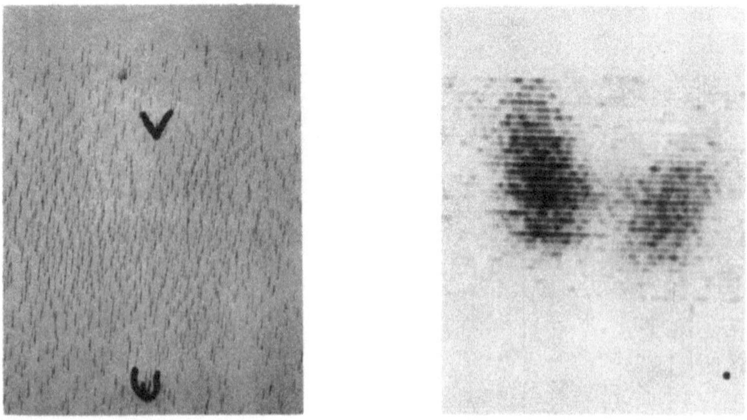

FIGURE 7. Thyroid imaging studies on a 51-year-old male with fever, painful thyroid enlargement, and symptoms of hyperthyroidism. The gland was modestly enlarged and very tender. Serum thyroxine, triiodothyronine, and thyroglobulin were all elevated. Thyroidal radioiodide uptake was below detectable levels and isotope scintigram (^{123}I) not satisfactory. Fluorescent scan was normal, however.

nodules much smaller than those palpable clinically in normal-sized thyroid glands. Such improvements would be even more helpful for nodules palpable with great difficulty in goitrous glands.

Application of these techniques to the world-wide problem of iodide deficiency goiter is expected, and may prove useful in the study of the efficacy of various dietary or parenteral methods for iodide supplementation for whole human populations. In more industrialized countries, where dietary iodide intake has greatly increased in recent decades, the future viability of isotopic thyroid imaging may come into question as the normal level of radioiodide uptake falls, and may result in even greater interest in FEA techniques.

Fluorescence excitation has already become the method of choice for thyroid imaging during pregnancy and lactation, in pediatric patients capable of cooperating during the imaging study, and in the large number of hospitalized patients undergoing thyroid imaging studies after administration of iodides as contrast media or medicinals. Such clinical uses will no doubt be exploited widely.

The finding that previously irradiated thyroid glands are sometimes iodide depleted, coupled with the observation that the iodide depleted gland in experimental animals is physiologically more sensitive to the effects of pituitary thyrotropin (TSH), may lead to changes in the understanding of radiation-induced thyroid disease and to changes in the clinical management of at least some of these patients, who are known to be at high risk for thyroid tumor development.

The lower levels of stable thyroidal iodine found in some elderly patients without other evidence of thyroid abnormality may lead to additional understanding of thyroid physiology in old age. For example, it may explain the common observation of a gradual enlargement of the gland with age, which is often seemingly unassociated with clinically demonstrable physiologic disturbance.

It also appears rational to study in a comprehensive manner the possible clinical usefulness of fluorescence excitation to quantitate stable glandular iodine in the course of prolonged antithyroid therapy of diffuse toxic goiter (Graves' disease) with antithyroid drugs. The untreated disease is characterized in most cases by increased total iodine stores, and the thiourylene drugs are known to interfere with thyroid hormone production by blockade of iodide organification. The demonstration of satisfactory organification blockage by fluorescence excitation appears feasible now, and may lead to some prognostic measure for selecting patients whose disease has achieved remission, or selecting those failing to remit for subsequent definitive treatment by surgery or sodium radioiodide. It now also appears feasible to relate the total glandular stable iodine content in Graves' disease to levels of glandular hormone on digests of thyroid tissue as determined in vitro after surgery; to study the interrelations of total glandular iodine level, glandular thyroxine, and triiodothyronine level; and to relate these quantities to the respective serum levels of thyroid hormones.

Study of the relationships of thyroidal stable iodine content to relative secretions of thyroxine and triiodothyronine and their secretion rates can be expected to clarify the roles of these two compounds in human physiology.

REFERENCES

1. Heedman, P.-A. and Jacobson, B., Thyroid iodine determined by X-ray spectrophotometry, *J. Clin. Endocr. Metabl.*, 24, 246, 1964.
2. Hoffer, P. B., Jones, W. B., Crawford, R. B., Beck, R., and Gottschalk, A., Fluorescent thyroid scanning: a new method of imaging the thyroid, *Radiology*, 90, 342, 1968.
3. Hoffer, P. B., Beckerman, C., Bowie, J., and Beck, R., Fluorescent Thyroid Scanning: What's New In Semiconductor Detectors in Medicine, *Oak Ridge, Tennessee*, UASEC CONF 730321, 1973, 238.
4. Hoffer, P. B., Fluorescent thyroid scanning, *Am. J. Roentgenol. Radium. Ther. Nucl. Med.*, 100, 721, 1969.
5. Okerlund, M. D., Sommers, J., Deconinck, F., Swann, S., and Kaufman, L., The clinical utility of simultaneous fluorescent thyroid scanning and intrathyroidal stable iodine quantitation, *Clin. Res.*, 25, 105A, 1977.
6. Palmer, D. W., Deconinck, F., Swann, S. J., Okerlund, M. D., and Highfill, R., Low cost intrathyroidal stable iodine quantitation with a fluorescent scanner, *Radiology*, 119, 733, 1976.
7. Okerlund, M. D., Palmer, D. W., Deconinck, F., Sommers, J., Swann, S., Lim, C., and Kaufman, L., Low cost intrathyroidal stable iodine quantitation with an improved fluorescent thyroid scanner, in *Medical Radionuclide Imaging*, Vol. I, International Atomic Energy Agency, Vienna, Austria, 1977, 485.
8. Patton, J. A., Hollifield, J. W., Brill, A. B., Lee, G. S., and Patton, D. D., Differentiation between malignant and benign thyroid nodules by fluorescent scanning, *J. Nucl. Med.*, 17, 17, 1976.
9. Chopra, I. J., Chopra, U., and Orgiazzi, J., Abnormalities of hypothalamo-hypophyseal axis in patients with Graves' ophthalmopathy, *J. Clin. Endocr. Metab.*, 37, 955, 1973.
10. Franco, P. S., Hershman, J. M., Haigler, E. D., Jr., and Pittman, J. A., Jr., Response to thyrotropin-releasing hormone compared with thyroid suppression tests in enthyroid Graves' disease, *Metabolism*, 22, 1357, 1973.
11. Okerlund, M. D., Sommers, J., Beckmann, A., Kaufman, L., Chuck, B., Ohara, J., Lim, C., Galante, M., and Hunt, T.,Studies on the Pathogenesis of Radiation-induced thyroid tumors. Program of the Annual Meeting of the American Thyroid Association, September, 1977. p. 1 (abstract).
12. Saenger, E. L., Silverman, F. N., Sterling, T. D., and Turner, M. E., Neoplasia following therapeutic irradiation for benign conditions in childhood, *Radiology*, 74, 889, 1960.
13. Winship, T. and Rosvall, R. V., Thyroid carcinoma in childhood, *Cancer*, 14, 734, 1961.
14. Duffy, B. J., Jr. and Fitzgerald, P. J., Cancer of the thyroid in children: a report of twenty-eight cases, *J. Clin. Endocr. Metab.*, 10, 1296, 1950.
15. Lindsay, S., Nichols, C. W., Jr., and Chaikoff, I. L., Induction of benign and malignant thyroid neoplasms in the rat. Induction of thyroid neoplasms by injection of 131I without or with feeding of diets containing propylthiouracil and/or dessicated thyroid, *Arch. Pathol.*, 81, 308, 1966.
16. Nadler, N. J., Mandavia, M., and Goldberg, M., The effect of hypophysectomy on the experimental production of thyroid neoplasms, *Cancer Res.*, 30, 1909, 1970.
17. Bray, G., Increased sensitivity of the thyroid in iodine-depleted rats to the goitrogenic effects of thyrotropin, *J. Clin. Invest.*, 47, 1640, 1968.

Index

INDEX

A

Accelerator, 49
  charged particle, 60
  electrostatic, 56
ADC, see Analog to digital convertor
Adsorption, 4, 14, 16—17
  coefficient, 17
  cross-section, 17
  edge, 10—11, 17—18
  photoelectric, 14
Alpha particle (He^{++}), 55
  20 MeV, 60
Amplifier, comparison of, 105
Analog to digital converter (ADC), 101
Atomic cross-section, 17
Atomic number (Z), 6
Atomic structure, 6—7
Atomic transition, 4, 7
Attenuation, 14, 16—17
Auger electron, 4, 12—13
  emission, 13
Avogardro's number, 17

B

Background, 98—99, 103—104, 111, 140
  dependence on tube voltage, 109
  limitation on FEA, 92—93
  variance vs. sample volume, 100
Barn, 17
Binding energy, 4
Blood, see also specific components
  cell survival study, 120
  cerebral, 124
  regional flow, measurement, 122—123
  whole, specific gravity, 120
Bragg reflection, 30
Brehmsstrahlung spectrum, 54
Bromine, quantitation, 104

C

Cardiac output, measurement of, 122—124
Cesium, 104
  blood disappearance rate, 119
Charged particle analysis, 60—63
Charge sensitive preamplifier, see Signal processing, input circuit preamplifier, charge sensitive
^{51}Chromium, blood disappearance rate, 119
Collimator, for FEA, 96—97
Compton scattering, 19—20
  energy shift as a function of scattering angle, 20
  minimization, 106
  relative probability vs. scattering angle, 21
Contrast agent, 129—130
  FEA study of, 130
    applications, 131—132
  kinetics, 129
Coster-Kronig transition, 12
Coulomb force, 6
Count rate, limitation on FEA, 94
Cyclotron, 56

D

Data processing, FEA electronics, 96—98
Dead time, 45
Deexcitation, 12
Detector, 94
  high-purity germanium, 32, 94—95
  lithium drifted silicon, 22, 23, 32, 94—95
    cross section, 33
    field distribution, 37
  semiconductor type, 30—32, 64
    background, 35—37
    design, 32
  efficiency, 34—35
    fabrication, 32—33
    signal, noise, and resolution, 33—34
Digital processing, 46

E

ECFV, see Extracellular fluid volume
Electron, 3, 6
  cross-section, 54
  excitation, 53—55
Electron beam
  microprobe, 63
  scanning electron microscope, see Scanning electron microscope
Electron volt (eV), 4
Energy-dispersive X-ray spectrometry, 4
Erythrocyte, volume, 117—119
Extracellular fluid vo lume (ECFV), measurement of, 116—117

F

Fano factor, 34
FEA, see Fluorescent-excitation analysis
Field-effect transistor, 37—39
Fine-focus electron-beam probe, see also Electron beam, 49
Fluorescence radiation-induced energy-dispersive analyzer (FRIEDA), see FRIEDA

Fluorescence yield (w), 12—14
atomic number, as a function of, 13
Fluorescent-excitation analysis (FEA), 4—5, 22—24
adsorption and attenuation, 14, 16—17
advantages and applications, 5—6
contrast media, study of, 130
applications, 131—133
in vitro, 91, 115
automated sample changer, 96, 98
blood cell survival, measurement, 120
cardiac output measurement, 122—124
cerebral blood flow and volume, measurement, 124
data processing electronics, 96—98
detector, 94—95
electronics, 95
extracellular fluid volume (ECFV), measurement of, 116—117
glomerular filtration rate (GFR), measurement, 121—122
plasma volume, measurement, 118—120
principles of, 92—94
pulmonary clearance of heavy metal powders, measurement, 124—125
red cell volume, measurement, 117—118
source for, 95—96
in vivo, 125—126
iodine containing sample, spectrum, 93
performance, 98—99
photoelectric adsorption, 17—18
polarized beam, 20—24, 109—113
equipment, 105—106
formulation for, 107—108
geometry, 105
improvement factor, 110-112
results, interpretation of, 108—109
target materials, 106—107
quantitation, 99—103
background, lowering of, 103—104
Compton suppression, 105
data rate increase, 104—105
scattering
coherent, 19
incoherent, 19—20
water sample, spectra of, 99
FRIEDA, 74
application of
hair, trace elements in, 80—81
metalloprotein, analysis of, 81—86
serum, trace element profile, 79—80
detection limit, 80
performance, 77—78, 81
sensitivity, 79
system, 75
X-ray probe, 76—77
spectrum of dried heart tissue excited with, 77

G

Germanium, high purity detector, see Detector, high purity Germanium
GFR, see Glomerular filtration rate
Glomerular filtration rate (GFR), estimation of, 121—122
Goiter, nodular, 156
Graves' disease, 151
thyroid imaging study, 152—154

H

Hair, trace elements in, 84—85
Hashimoto's thyroiditis, 155—156
Heavy charged particle
beam, 49
excitation, 55—57
ionization cross-section, 55
microprobe, 56
Hyperthyroidism, thyroid imaging study, 155, 157—159

I

^{125}I-iothalamate, blood disappearance rate, 121—122
Image contrast, 141
Internal conversion, 14
Intracranial pressure, measurement of, 125
Iodine
determination, in vivo, 132—133, 150
fluorescence yield, 22
intrathyroidal, 139, 145—147, 150
Iodipamide, whale blood iodine following injection, 129—130
Iopanoate, biliary excretion, 131—132
Isotope, 6

K

KeVex® scanner, 143
calibration curve, 146
Kilo-electron volt (keV), 4
K shell, 7
vacancy production
electron excitation, 54
photon excitation, 51
proton excitation, 56

L

Linear adsorption coefficient (N), 16
Lithium-drifted silicon detector, see Detector, lithium-drifted silicon
L shell, 7
Lung, clearance of heavy metal powders, 124—125

M

Manganese, X-ray spectrum, 31
Mass adsorption coefficient (Nm), 17
Mass number (A), 6
Menke's Kinky Hair Syndrome (MKHS), 79
  trace element profile, 82—83
Metalloprotein, 81—86
  two step study, schematic presentation, 86
MKHS, see Menke's Kinky Hair Syndrome
Multichannel analyzer, 22

N

Neutron (N), 6
Nuclear charge, 6—7
Nucleus, 3, 6

O

Oak Ridge Technical Enterprises Corp. scanner, 143
Orbital, angular momentum, 7

P

Particle induced X-ray emission analysis (PIXE), 60
Pauli exclusion principle, 6
Periodic Table of the Elements, 6, 25
PHA, see Pulse height analyzer
Phenobarbital, effect on biliary excretion, 131
Photoelectric cross-section, 53
Photon excitation, 50—53
PIXE, see Particle induced X-ray emission analysis
Plasma, volume, 118—120
Princeton Gamma-Tech Scanner, 143
Principal quantum number, see Quantum number, principal quantum number
Proton (H^+), 6, 55
  3 MeV, 60
  microprobe, 64—65
Pulse-height analyzer (PHA), 4, 96
  continuum photon fluorescence, 53
  drift, 102
  electron excitation, 55
  monoenergetic photon fluorescence, 52
  proton excitation, 57
Pulse shaping, effect on resolution, 41—43
Pulsed-excitation technique, 45—46
Pulsed-light feedback, 40—41

Q

Quantum number, 6—7
Quantum theory, 3

R

Rabbit, plasma volume, 120
Radiation
  adsorption, see Adsorption
  scattering, see Scattering
Radiationless transition, 12
Radioisotope
  excitation, 49, 59—60
    photon excitation, 50
Radio-opaque contrast media, see Contrast agent
Rayleigh scattering, 19
Red cell, see Erythrocyte
Resolution, 43

S

Satellite line, 13
Scanner, fluorescent, thyroid imaging, see Thyroid, fluorescent scanning
Scanning electron microscope (SEM), 64
Scattering, 4, 52
  coherent, see Rayleigh scattering
  incoherent, see Compton scattering
  radiation, 14
Secondary fluorescence, 57
Self adsorption, 14
SEM, see Scanning electron microscope
Sensitivity, fluorescent scanner, 139
Serum, trace elements in, 73—74, 79—80
Shell, 6
Signal processing
  amplifier
    gain, 41
    linearity, 41
    pulse shaping, effect on resolution, 41—43
  digital processing, 46
  input-circuit preamplifier
    charge sensitive, 39—40
    field-effect transistor, 37—39
    pulsed-light feedback, 40—41
  pulsed-excitation technique, 45—46
  system considerations, 44—45
Sodium iopanoate, biliary excretion, 131—132
Spatial resolution (R), 139
Spectrometer
  crystal, 30

semiconductor type, 38
Subshell, 6
Superheavy element, 6
Synchrotron radiation, 65

T

Thyroid
fluorescent scanning, 139, 149—160
calibration curve, 146
formulation for, 139—142
instrument, schematic representation, 140
results, 142—145
sensitivity, 143, 144
spatial resolution, 144
nodule
functioning, 151—152
nonfunctioning, 151
radiation induced disease, 156—157
Trace element, 71—73
comparison between seawater and normal human serum, 72
serum, level in, 73—74
tissue, level in, 74
Transition, see Atomic transition

W

Wavelength-dispersive spectrometry, 4

X

X-ray
contrast agent, see Contrast agent
energy, 10—11
energy-dispersive spectrometry, see Energy-dispersive X-ray spectrometry
escape peak, 35—36
fluorescence excitation, 50
electron excitation, 53—55
energy dispersive, 50
heavy charged particle, 55—57
photon excitation, 50—53
wavelength dispersive, 30
fluorescence yield, 12—14
intensity, 12—13
K-series
emission intensity, 15
energy, 9
low energy, detection of, 65—66
L-series, 7
emission intensity
energy, 10
M-series, 8
nomenclature, 7—12
polarized beam, 49, 65
secondary target production, 58
spectrometer, 44
tube, 49
fluorescence excitation, 57—59
photon excitation, 50

Z

Zirconium
X-ray spectrum, 36